Springer Tracts in Advanced Robotics

Volume 9

Editors: Bruno Siciliano · Oussama Khatib · Frans Groen

Springer

Berlin
Heidelberg
New York
Hong Kong
London
Milan
Paris
Tokyo

Katsu Yamane

Simulating and Generating Motions of Human Figures

With 73 Figures

Springer

Professor Bruno Siciliano, Dipartimento di Informatica e Sistemistica, Università degli Studi di Napoli Federico II, Via Claudio 21, 80125 Napoli, Italy, email: siciliano@unina.it

Professor Oussama Khatib, Robotics Laboratory, Department of Computer Science, Stanford University, Stanford, CA 94305-9010, USA, email: khatib@cs.stanford.edu

Professor Frans Groen, Department of Computer Science, Universiteit van Amsterdam, Kruislaan 403, 1098 SJ Amsterdam, The Netherlands, email: groen@science.uva.nl

STAR (Springer Tracts in Advanced Robotics) has been promoted under the auspices of EURON (European Robotics Research Network)

Author

Dr. Katsu Yamane
Department of Mechano-Informatics
Graduate School of Information Science and Technology
University of Tokyo
7-3-1, Hongo, Bukyo-ku
113-8656 Tokyo, Japan
yamane@ynl.t.u-tokyo.ac.jp

ISSN 1610-7438

ISBN 3-540-20317-6 Springer-Verlag Berlin Heidelberg New York

Cataloging-in-Publication Data applied for
A catalog record for this book is available from the Library of Congress.
Bibliographic information published by Die Deutsche Bibliothek
Die Deutsche Bibliothek lists this publication in the Deutsche Nationalbibliografie; detailed bibliographic data is available in the Internet at <http://dnb.ddb.de>.

Springer-Verlag is a part of Springer Science+Business Media

springeronline.com

Typesetting: Digital data supplied by author.
Data-conversion and production: PTP-Berlin Protago-TeX-Production GmbH, Berlin
Cover-Design: design & production GmbH, Heidelberg
Printed on acid-free paper 62/3020Yu - 5 4 3 2 1 0

To Noriko, who supported me in all non-research issues

Foreword

At the dawn of the new millennium, robotics is undergoing a major transformation in scope and dimension. From a largely dominant industrial focus, robotics is rapidly expanding into the challenges of unstructured environments. Interacting with, assisting, serving, and exploring with humans, the emerging robots will increasingly touch people and their lives.

The goal of the new series of Springer Tracts in Advanced Robotics (STAR) is to bring, in a timely fashion, the latest advances and developments in robotics on the basis of their significance and quality. It is our hope that the wider dissemination of research developments will stimulate more exchanges and collaborations among the research community and contribute to further advancement of this rapidly growing field.

The monograph written by Katsu Yamane is an evolution of the Author's Ph.D. dissertation. The work builds upon a rising interest in human figures which is inspired by recent research on humanoid robotics and computer animation. A foundation for dynamic modeling of complex kinematic chains is established and original methods for interactive generation of human figure motions are developed. Key issues of the approach are physical consistency and interactivity, which are critically examined in the course of devising efficient computational algorithms for real-time dynamics simulation, as well as frameworks for interactive and intuitive graphical animation.

One of the first focused books in the fascinating field of humanoids and human-centered robotics, this title constitutes a very fine addition to the series!

Naples, Italy
August 2003

Bruno Siciliano
STAR Editor

Preface

This book is a collection of original algorithms to simulate, analyze, and generate motions of human figures, all focusing on realtime and interactive computation. The target readers are graduate students and researchers in humanoid robotics and computer animation fields who are looking for advanced techniques for interactively simulating and generating motions of human figures. Therefore, knowledge on basic kinematics and dynamics computations of robot manipulators is a prerequisite to understand the algorithms described in this book.

Compared to conventional robotic manipulators, human figures are characterized by their complexity and instability. One of the significant features of human figures is that the link connectivity changes during a motion, namely, human figures are *structure-varying* kinematic chains and they form complex linkages including closed kinematic chains. Another feature is that a motion is not always feasible for a specific human figure due to its instability, or more formally, under-actuatedness. *Physical consistency* is the condition that a motion is feasible for the target human figure with a choice of internal (joint) forces. This condition is especially important if we are controlling a physical robot, but it is also valuable for computer animation because it can add physical naturalness to a motion.

On the other hand, most of the applications of human figures require high interactivity in motion generation. Humanoid robots, for example, are expected to work in daily environment with naive human co-workers. They have to generate their motions in real time in response to the changes of environments or high-level inputs from human co-workers. Because computation of full dynamics of such complicated kinematic chains requires heavy computational load with currently available algorithms, interactivity of motion generation has not been discussed very much at physics or motion level.

This book tries to address these problems by developing efficient techniques for the dynamics computation and interactive frameworks for the motion generation of human figures.

This book is based on my Ph.D. dissertation written under the supervision of Professor Yoshihiko Nakamura at Department of Mechano-Informatics, the University of Tokyo. Professor Nakamura was the supervisor throughout my graduate school study. His continuing support, encouragement, and stimulation were essential to the completion of the thesis. Discussion with him was always a source of inspiration.

I would also like to thank the following members of Nakamura & Okada Laboratory for developing the motion capture system, being subjects of motion capture sessions, and providing motion captured data used in the examples: Shinji Hara, Shin'ichiro Hoshino, Kazutaka Kurihara, Shoji Okamoto, and Ichiro Suzuki.

The research was supported by many projects and funds. First of all, I would like to gratefully acknowledge the support from the Japan Society for the Promotion of Science as a research fellow. The fundamental fund for the research activities was provided by the CREST program of the Japan Science and Technology Corporation through "Development of Machines with Brain-Like Information Processing through Dynamical Connections of Autonomous Behavioral Primitives" (PI: Y. Nakamura, University of Tokyo).

The results presented in Chapters 2 and 3 were obtained in the collaborative project with Fujitsu Laboratories, Ltd., under the support by the Advanced Software Enrichment Project of the Information-technology Promotion Agency (IPA), Japan. IPA also supported our project with Sega Corporation "Development of New Application for Motion Generation Based on Control Techniques for Humanoid Robots" (PI: H. Imagawa, Sega Corporation) from which the results in Chapter 8 were derived. The simulation methods in Chapters 4–6 were developed under the Humanoid Robotics Project directed by New Energy and Industrial Technology Development Organization (NEDO), Japan.

The work in Chapter 10 was done with Professor Jessica K. Hodgins while I was a postdoctoral fellow at the Robotics Institute of Carnegie Mellon University, and was supported in part by NSF 0196089 and 0196217. The hardware was constructed by Dr. H. Benjamin Brown. Rory Macey and Justin Macey assisted us in the motion capture sessions for capturing human and marionette motions.

I wish to thank Professor Bruno Siciliano and Dr. Thomas Ditzinger for their patience and support during the preparation of the manuscript.

Finally, I would like to thank my family, especially my wife Noriko Yamane, for their support in all non-technical issues.

Tokyo, Japan
August 2003

Katsu Yamane
University of Tokyo

Contents

Part III Conclusion

1

Introduction

1.1 Background

This book focuses on two issues related to motions of human figures: realtime dynamics computation and interactive motion generation. In spite of the growing interest in human figures as both physical robots and virtual characters, standard algorithms and tools for their kinematics and dynamics computation have not been investigated very much. Researchers have to develop their own tools by themselves or manage to apply existing ones designed for conventional robotic manipulators. The algorithms presented in this book are expected to provide fundamental computational tools for analyzing, simulating and controlling motions of human figures.

Human figures are defined as mechanical models of human body. Applications of such models include humanoid robots and character animation in computer graphics (CG). A typical human figure would consist of two legs, two arms, a head, and a trunk, having enough mechanical ability to perform motions almost equivalent to those of a real human. The total degrees of freedom (DOF) varies from 20 to 50 or more, depending upon how precise the model should be to satisfy the objective for which the model is used. The links are usually connected by mechanical joints, e.g. rotational, spherical, or sometimes prismatic ones. Anatomical human models using muscles and tendons would also have wide range of applications such as orthopedic surgery planning, rehabilitation, sport training, and so forth.

Obviously, human figures have completely different properties from conventional robotic systems. First of all, they contain many DOF which require huge amount of computation for controlling and simulating their motions. Secondly, most human figures have free-flying base and no link is fixed to the inertial frame. As a result, control and motion generation of human figures can be much more difficult. Thirdly, because many human motions make use of collisions and contacts between the body and the environment, we have to consider these phenomena in simulating and generating motions of human figures. However, it is known that collisions and contacts are difficult to model due to their high frequency and unilateral constraints. Finally, most of the practical motions of human figures involve different closed and/or open kinematic chains and the link connectivity may change during a motion. In walking, for example, the link connectivity of our body changes among left-leg support, right-leg support and both-leg support phases.

In spite of the difficulties, efficient dynamics computation and motion generation techniques for human figures are in urgent need. We can find two major backgrounds for rising interest in human figures: humanoid robotics and computer animation.

After the sensational debut of the Honda humanoid robot P3 in 1997, the Japanese government launched the five-year project called Humanoid Robotics Project (HRP) in 1998. The first two years were dedicated to the development of three platforms—hardware, software (simulator), and cockpit—which were then applied to a number of applications in the next three years. Many companies also followed by revealing various types of humanoid robots including bipeds. Although they can perform a number of interesting motions, they are still not autonomous in the sense that each new motion has to be carefully programmed and tested before it is actually performed. This is probably the reason why current humanoid robots are mainly used for entertainment purposes. In order to work in daily environments, humanoid robots have to update their motions in real time according to the current environment and input from the user.

Recently computer graphics is widely used in areas including films and games where the quality of human and/or animal characters' motions are critical. However, today's computer animation software packages require knowledge and skill of extremely high level from animators to create realistic motions of human and animal figures. Eagerly demanded is a tool or a computational algorithm that is capable of generating motions of human figures easily without any knowledge about physics, biology or computer science, just as choreographing a real dancer.

Considering the potential application areas and difficulty in control of human figures, we find two key issues in motion generation: physical consistency and interactivity.

A randomly generated motion is not always feasible even for an ideal human figure when no link is fixed to the inertial frame. *Physical consistency* represents whether a motion is physically feasible for a specific human figure with a choice of internal forces. Many factors can cause physical inconsistency: infeasible contact forces, infeasible joint torques, angles or velocities, self collisions, or collisions with the environment. Physical consistency is the primary requirement for motions of humanoid robots to avoid unexpected motions and realize safe operation, while in computer animation it is not necessarily required. In some computer graphics applications, however, satisfying physical consistency would help animators to add physical reality to motions.

Interactivity is an important issue in both humanoid robots and computer animation. Humanoid robots, expected to work in daily environment, have to change its motion according to the environment and interactions with human co-workers, who are usually naive and can only give high-level commands to the robot. In computer animation, interactivity and intuitive interface are essential requirements for animation tools because animators are not always experts in physics or mathematics.

Unfortunately, the two issues conflict with each other: the physical consistency requires the computation of time-consuming full dynamics of complicated human figures, while interactivity requires realtime motion generation with limited information. This book tries to cope with the two issues by developing efficient algorithms for realtime dynamics computation and frameworks for interactive motion generation. The following two subsections discuss the specific difficulties in dynamics computation and motion generation.

1.1.1 Realtime Dynamics Computation

Inverse and forward dynamics computation are important tools for analyzing and simulating motions of mechanical systems including human figures. They also serve as the basic tools for generating motions taking the dynamics into account. Efficient algorithms for these computations would accelerate the research, design and development of humanoid hardware, controllers and motion generation techniques.

Because human figures consist of mechanical joints as most robotic systems do, it is indeed possible to apply conventional techniques for the dynamics computation of kinematic chains to human figures. Compared to robotic manipulators for which most dynamics computation algorithms are designed, however, the motions of human figures have the following four major properties:

(1) The system usually contains many DOF. A typical human figure would have at least 20 DOF and sometimes more than 50, even ignoring the fingers. Due to the complexity, they require much larger computational cost for kinematics and dynamics than conventional robotic manipulators.
(2) The link connectivity may change during a motion from an open kinematic chain to a closed one and vice versa, by catching or releasing an object with the hands. Such systems are called *structure-varying* kinematic chains in this book. Most of the conventional algorithms for dynamics computation require at least a reconstruction of the equation of motion, and some are even inherently impossible, to handle this type of kinematic chains.
(3) Human figures frequently form complicated closed kinematic chains by holding objects or coming into contact with the environment. For example, simply holding a bar with the both hands can generate a closed kinematic chain. Realtime dynamics computation of general closed kinematic chains is still a challenging problem.
(4) Human figures are often subject to collisions and contacts with the environment or itself during their motion. In some cases, humans even make use of such phenomena to achieve agile motions. Unfortunately, collisions and contacts are known to be very difficult to handle in computer simulation as well as in hardware experiments. The difficulty in computer simulation comes from their high frequency and unilateral constraints. In spite of continuing efforts, fast and stable collision/contact simulation is still an open research issue.

Due to these differences, applying techniques for conventional robotic manipulators is not always effective. This book describes several algorithms that are capable of handling structure-varying kinematic chains and collisions/contacts efficiently.

1.1.2 Interactive Motion Generation

Physical consistency and interactivity are the two key issues in motion generation of human figures. Interactive and autonomous motion generation of human figures is investigated in both humanoid and computer animation fields.

A physically consistent motion should satisfy various physical constraints such as

- the total angular momentum should be constant during free-flying motions
- the zero moment point (ZMP) should stay in the contact area

as well as other constraints depending on the design of a specific robot such as joint torque limits, joint angle limits and so on. It is true that online controllers, which is out of the scope of this book, are also essential for controlling a real robot. Ensuring physical feasibility of the reference motion, however, would increase the stability of the resulting motions and facilitate the design of online controllers. In fact, most online controllers for humanoid robots aim at compensating unexpected modeling errors and disturbances, assuming that the predefined motion patterns are physically feasible for the ideal model of the target robot.

In humanoid robotics, where stability is the primary requirement, physical consistency of motions has been given the higher priority and interactivity has mostly been discussed on the behavior level as a problem of modeling human-like intelligence and information processing. In addition, motion patterns generated by existing schemes tend to be artificial, *robot-like* motions for simplicity and safety. However, as they come to work in complicated and dynamic environment, they should generate human-like motions interactively in response to changes in the environment, inputs from the user and so forth, not just following pre-recorded motion patterns. Interactivity would thus become another important issue for motion generation of humanoids.

In computer animation, interactivity is obviously essential for some applications such as games. Even in creating films, animators would prefer manipulating the characters interactively using intuitive devices to tuning parameters and waiting for a couple of minutes to get the output. However, creating realistic motion of human characters in computer animation still relies heavily on an animator's skill or motion capture techniques. Moreover, even if a motion clip was created through hard work, it is difficult to modify it to reuse in another scene or for a different character. Physical consistency also plays an important role in some applications of computer animation because animators can obtain realistic motions without any knowledge about physics.

1.2 Outline

The goals of this book are to establish a foundation for realtime dynamics computation of structure-varying kinematic chains including human figures, and to develop methods for interactive generation of a variety of motions of human figures. The

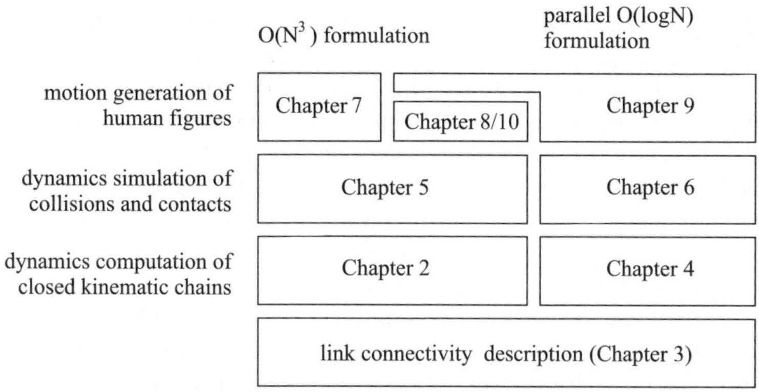

Fig. 1.1. Relationships between the methods presented in each chapter.

first half, Chapters 2 to 6, describes different approaches to simulating the motion of structure-varying kinematic chains and contacts. The latter half, Chapters 7 to 10, describes interactive motion generation techniques based on the algorithms presented in the first half. Finally the concluding remarks are to be addressed in Chapter 11.

The relationships between the methods presented in this book are illustrated in Fig. 1.1. They are divided into three categories according to the purposes and two categories according to the formulation of dynamics, all based on the fundamental link connectivity description technique in Chapter 3. The three purposes include basic dynamics computation for closed kinematic chains, collision/contact simulation, and motion generation. The two formulations are characterized by their asymptotic complexity for forward dynamics—$O(N^3)$ of those in Chapters 2 and 5, and parallel $O(\log N)$ of those in Chapters 4 and 6.

The relationships of motion generation techniques in Chapters 8 and 9 with dynamics computation algorithms are somewhat complicated. The pin-and-drag interface presented in Chapter 8 considers only the kinematic constraints which are expressed in the same way as in the $O(N^3)$ formulation. Chapter 9 presents a motion generation technique based on the $O(\log N)$ formulation of dynamics as well as the pin-and-drag interface developed in Chapter 8. Chapter 10 presents an application of the method in Chapter 8 to robot control.

The following subsections briefly summarize the algorithm presented in each chapter.

1.2.1 Realtime Dynamics Computation

In Chapter 2, a new method for inverse and forward dynamics of general closed kinematic chains is to be presented. The method is based on the efficient algorithm for inverse dynamics computation of closed kinematic chains developed by Nakamura and Ghodoussi [63]. The inverse dynamics is first computed for the virtual open kinematic chain obtained by virtually cutting several joints in the original closed kinematic chain. The joint torques are then transformed to those of the actuators by mean of the Jacobian matrix of the angles of the actuated joints with respect to those of the virtual open kinematic chain. Although the method is potentially applicable to any closed kinematic chains, general algorithm to compute the Jacobian matrix was not known.

Chapter 2 presents a general method to compute the Jacobian matrix used for force transformation and thus enables inverse dynamics computation of general closed kinematic chains. The forward dynamics algorithm is then derived based on unit vector method [96] which repeats inverse dynamics computation to obtain the inertial matrix in the joint space. The resulting method can compute the inverse and forward dynamics of any closed kinematic chains more efficiently than conventional methods in commercial softwares [31]. However, since its asymptotic complexity is $O(N^3)$, the method is still not efficient enough for large systems such as crowd or precise anatomical human models.

In Chapter 3, the method is extended to structure-varying kinematic chains by introducing the link connectivity description using *pointers* and *virtual links*. The advantage of our approach is that, thanks to its close relationship with the most common programming languages such as C/C++ and Java, it is the most effective way for implementing dynamics computation algorithms, especially recursive ones such as Newton-Euler Formulation. Moreover, internal description is easily and rapidly changed by only adding or removing virtual links in accordance with the structural changes. Combined with the dynamics computation algorithm applicable to general open and closed kinematic chains, it enables efficient dynamics simulation of structure-varying kinematic chains. Computation of the velocity boundary condition after link connection is also discussed.

In Chapter 4, an improved forward dynamics computation method is to be presented. Although it is based on the same scheme as in the previous method, the asymptotic complexity is as small as $O(N)$ for serial computation, which is comparable to the most efficient algorithms currently available. In addition, parallel computation on $O(N)$ processors reduces the complexity down to $O(\log N)$ without changing the algorithm. One of the features of the method is that the parallelism can be tuned for any number of processors without modifying the program, which is a great advantage in handling structure-varying kinematic chains.

The method involves two procedures which are physically interpreted as assembling and disassembling the target kinematic chain by adding or removing joints one by one. The $O(N)$ complexity is achieved by reducing the dynamics of each intermediate kinematic chain into the relationships of forces applied to its end links and their accelerations. If the number of end links of each intermediate chain is bounded, which is the case for all open kinematic chains and most of closed ones, the amount of computation for each assembly and disassembly procedure is also bounded. The total computational load is, therefore, proportional to the number of joints.

Parallel computation is realized by optimizing the assembly order such that the intermediate chains become independent of each others as much as possible. If the chains constructed by two assembly procedures are independent, the computations can be run in parallel on two processors, taking only the same CPU time as a single assembly. Because the assembly order uniquely determines the parallelism, it is relatively easy to optimize the parallelism of the dynamics computation of an arbitrary kinematic chain for parallel computation on any number of processors.

In Chapter 5, a collision/contact simulation method based on the equation of motion developed in Chapter 2 is presented. It is basically a rigid-body contact model in the sense that it does not use the deformation of the links in contact. However, unlike most rigid contact models which employ time-consuming optimization algorithms, an iterative trial-and-error procedure is applied to find the contact force and constraint conditions that satisfy the unilateral constraints. The advantages of rigid contact models are that we need almost no parameter tuning and that we can use relatively large time steps for integration, although each step may take long time due to the optimization process. By simplifying the computation for optimization, the method realizes fast and stable simulation of collisions and contacts.

The method presented in Chapter 6, on the other hand, is based on soft contact model and enables large time steps by partially adapting implicit integration [11]. Implicit integration is a common technique for simulating systems with high stiffness such as cloth and particles. Its main idea is to use the force at the next time step rather than the current to obtain stable and realistic results with large time step. The simulation system combining the soft contact model and the dynamics computation in Chapter 4 takes only a few times longer than real time on a PC to simulate the motions of articulated chains with collisions and contacts, including collision detection, graphics representation and controllers.

1.2.2 Interactive Motion Generation

Chapters 7 to 10 present various methods for generating human-like motions of human figures interactively.

The first method described in Chapter 7, called *dynamics filter*, considers the full dynamics of human figure to generate a physically consistent motion from a reference motion that might be physically inconsistent. The reference motion can be any captured, hand-made, or numerically generated one. The advantage of the method is that it only requires time-local information of the reference motion. We can therefore give a reference motion generated online according to, for example, interactions with the user.

The dynamics filter is composed of two blocks: controller and optimizer. The controller computes the *desired* accelerations from the reference motion and current state, which might not be physically feasible. The desired accelerations are then input to the optimizer which computes the *actual* accelerations that is close to the desired but satisfy physical consistency. The actual accelerations are integrated to yield the filtered motion.

The second method presented in Chapter 8 is mainly intended for computer animation in the sense that it only considers kinematic constraints. Its main purpose is to provide animators with a simple and intuitive interface to generate whole-body motions of human figures. It allows the users to generate motions by simply dragging a link with any number of links pinned, thus the interface is called *pin and drag*.

Although the method basically does not require any reference motions, it is naturally extended to editing pre-recorded motion interactively.

The key technique used here is the enhanced inverse kinematics algorithm that is capable of handling various types of constraints at the same time avoiding singularity. Users can attach any number of constraints such as joint motion range, desired joint values in addition to the pin and drag constraints, without worrying about the singularity and inconsistency between the constraints. Singularity-robust (SR) inverse [64] plays an important role in avoiding singularity.

The third method, presented in Chapter 9, combines the efficient dynamics computation in Chapter 4 and the interface in Chapter 8 to generate physically feasible motions that satisfies the kinematic constraints input through the interface while satisfying the physical constraints at the same time. The method utilizes the inertial matrices that relates the forces and the accelerations of the end links of human figure, called the operational space inertia matrix [48]. Used with the pin-and-drag interface, these matrices characterize the relationships between the forces and accelerations of the pins and dragged links and enable us to ignore the effect of the rest of the body.

Chapter 10 applies the pin-and-drag interface in Chapter 8 to controlling a motor-driven marionette using human motion capture data as a reference. The marker positions measured by an optical motion capture system can be regarded as mutiple *drag* points. The joint and motor angles are computed from these marker position as well as the new class of constraint that maintains the length of the strings. This chapter also deals with hardware-specific problems such as mapping the marker positions into the marionette's workspace and preventing swings.

Part I

Dynamics Computation of Structure-Varying Kinematic Chains

2

Inverse and Forward Dynamics of General Closed Kinematic Chains

2.1 Introduction

This chapter describes the inverse and forward dynamics algorithms for general closed kinematic chains.

The inverse dynamics computation is based on the approach proposed by Nakamura and Ghodoussi [63]. In this approach, first the inverse dynamics is computed for the virtual open kinematic chain obtained by virtually cutting several joints of the original closed chain, assuming all joints except for the cut ones are actuated and the open chain performs exactly the same motion as the original one. Efficient algorithms such as Newton-Euler formulation [71] can be applied here. Next, the joint torques of the open kinematic chain are transformed into those of the actuated joints using the Jacobian matrix of the angles of the actuated joints with respect to those of the open kinematic chain, based on the equation derived from d'Alembert's principle.

The difficulty of inverse dynamics computation of closed kinematic chains comes from the unknown loop constraint forces. The most common approach is to apply Lagrange multipliers to take care of the constraint forces [42]. The method presented in [63] is more computationally efficient because it avoids explicit computation of constraint forces by making use of the joint torques of the virtual open kinematic chain.

The problem of their method is that there is no general way to compute the Jacobian matrix for force transformation. For some simple closed kinematic chains such as planar mechanisms, the Jacobian matrix is easily computed utilizing a geometric insight, sometimes even being constant. However, the computation is not that simple for general closed kinematic chains, especially for spatial structures.

The main contribution of this chapter is the general algorithm for computing the Jacobian matrix. It automatically computes the degrees of freedom and selects the

---

This chapter was adapted from, by permission, Y. Nakamura and K. Yamane, "Dynamics Computation of Structure-Varying Kinematic Chains and Its Application to Human Figures," IEEE Transactions on Robotics and Automation, vol.16, no.2, pp.124–134, 2000.

generalized coordinates of the system from the Jacobian matrix of the constraint point with respect to the joint angles in a simple and intuitive way without any assumption on the structure. Combining the algorithm presented here with the method in [63], we can efficiently compute the inverse dynamics of general closed kinematic chains.

The forward dynamics algorithm is derived based on unit vector method [96], where the joint-space inertial matrix is computed by repeating the inverse dynamics computation. The inertial matrix is then inverted to compute the generalized accelerations. Although the overall asymptotic complexity of forward dynamics computation is $O(N^3)$ where N denotes the degrees of freedom, the coefficient of the most significant term is much smaller than existing methods for forward dynamics computation of general closed kinematic chains.

This chapter is organized as follows. After briefly reviewing the previous work on the dynamics computation of closed kinematic chains in Section 2.2, we introduce the generalized coordinates of closed kinematic chains by extending the discussion in [63]. We then present the method to compute the Jacobian matrix, used for computing the inverse dynamics of closed kinematic chains, in Section 2.3. Section 2.6 summarizes the inverse and forward dynamics algorithms for general closed kinematic chains. Section 2.7 proposes to utilize multi-DOF joints to reduce the computational cost. Finally, Section 2.8 shows a simulation example involving a complicated closed chain, as well as a simple numerical example that demonstrates how our algorithm works.

2.2 Related Work

Many efficient methods have been developed for open kinematic chains from relatively early days such as Newton-Euler Formulation [71, 52, 69] for the inverse dynamics and Composite-Rigid-Body Method [96], Articulated-Body Method [22] and other efficient recursive algorithms [7, 89] for the forward dynamics. Efficient dynamics computation of closed kinematic chains, on the other hand, is much more difficult and still under investigation [22, 8]. The main problem is that we have to cope with the unknown loop constraint force in some way. In current commercial software [31], all constraints are solved simultaneously using global coordinates and a large equation of motion. The most common alternative to this approach is to use Lagrange multipliers to compute the unknown constraint forces [42]. Baraff [12] showed that making use of the sparsity of the coefficient matrix of the equation of motion greatly reduces the computational load.

2.3 Generalized Coordinates of Closed Kinematic Chains

Consider a closed kinematic chain in Fig. 2.1. Let N_J be the total number of joints in the chain, $\boldsymbol{\theta}_J \in \boldsymbol{R}^{N_J}$ the whole joint angles, N_A the number of actuated joints, $\boldsymbol{\theta}_A \in \boldsymbol{R}^{N_A}$ the actuated joints and $\boldsymbol{\tau}_A \in \boldsymbol{R}^{N_A}$ the actuator torques. In this section,

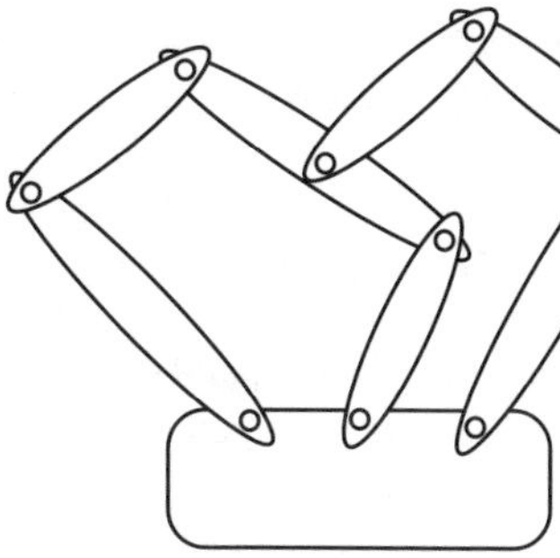

Fig. 2.1. A closed kinematic chain.

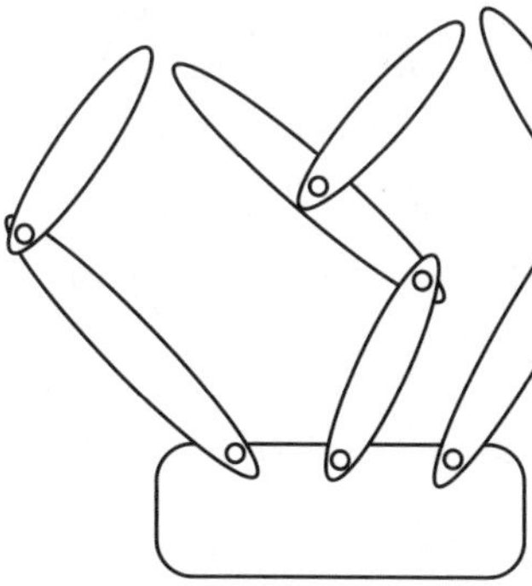

Fig. 2.2. The tree-structure open kinematic chain associated with the closed kinematic chain in Fig. 2.1.

we assume that the mechanism has rotational or translational joints of single-degree-of-freedom for simplicity sake. Introducing multi-degree-of-freedom joints requires no essential modification to the algorithm, and will be discussed later in Section 2.7.

Suppose that the closed chain is virtually cut at some joints and forms a tree-structure open kinematic chain in Fig. 2.2. Let N_O be the number of joints in the tree-structure chain, $\boldsymbol{\theta}_O \in \boldsymbol{R}^{N_O}$ the joint angles and $\boldsymbol{\tau}_O \in \boldsymbol{R}^{N_O}$ the joint torques. We assume at this moment that all joints in the tree structure, including those unactuated in the original closed chain, are actuated.

Suppose that the tree structure makes the same motion as the original closed chain without force or moment interaction at the virtually cut joints. The joint torques $\boldsymbol{\tau}_O$ required to generate the motion is computed by recursive inverse dynamics algorithms for open kinematic chains [96, 52, 69]. Note that nonzero values may be obtained for the elements of $\boldsymbol{\theta}_O$ corresponding to the unactuated joints in the original closed kinematic chain.

Let the original closed kinematic chain have N_F degrees of freedom, $\boldsymbol{\theta}_G \in \boldsymbol{R}^{N_F}$ be the generalized coordinates that describe the mobility of the closed kinematic chain, and $\boldsymbol{\tau}_G$ be the generalized force. We can form $\boldsymbol{\theta}_G$ by selecting appropriate

N_F joints from $\boldsymbol{\theta}_J$, for instance. Since the generalized coordinates determine the motion of the whole mechanism, $\boldsymbol{\theta}_A$ and $\boldsymbol{\theta}_O$ can be written as follows:

$$\boldsymbol{\theta}_O = \boldsymbol{\theta}_O(\boldsymbol{\theta}_G) \tag{2.1}$$

$$\boldsymbol{\theta}_A = \boldsymbol{\theta}_A(\boldsymbol{\theta}_G). \tag{2.2}$$

From Eq.(2.1), the d'Alembert's principle, and the principle of virtual work, the joint torques of the tree structure $\boldsymbol{\tau}_O$ and the generalized forces $\boldsymbol{\tau}_G$ satisfy the following equation [63]:

$$\boldsymbol{\tau}_G^T \delta\boldsymbol{\theta}_G = \boldsymbol{\tau}_O^T \delta\boldsymbol{\theta}_O = \boldsymbol{\tau}_O^T \boldsymbol{W} \delta\boldsymbol{\theta}_G \tag{2.3}$$

where

$$\boldsymbol{W} \triangleq \frac{\partial \boldsymbol{\theta}_O}{\partial \boldsymbol{\theta}_G} \in \boldsymbol{R}^{N_O \times N_F}. \tag{2.4}$$

$\delta\boldsymbol{\theta}_O$ and $\delta\boldsymbol{\theta}_G$ are the virtual displacements of $\boldsymbol{\theta}_O$ and $\boldsymbol{\theta}_G$, respectively. Similarly, Eq.(2.2) and the principle of virtual work yield

$$\boldsymbol{\tau}_G^T \delta\boldsymbol{\theta}_G = \boldsymbol{\tau}_A^T \delta\boldsymbol{\theta}_A = \boldsymbol{\tau}_A^T \boldsymbol{S} \delta\boldsymbol{\theta}_G \tag{2.5}$$

where

$$\boldsymbol{S} \triangleq \frac{\partial \boldsymbol{\theta}_A}{\partial \boldsymbol{\theta}_G} \in \boldsymbol{R}^{N_A \times N_F}. \tag{2.6}$$

$\delta\boldsymbol{\theta}_A$ is the virtual displacement of $\boldsymbol{\theta}_A$. Since Eqs.(2.3) and (2.5) hold for any $\delta\boldsymbol{\theta}_G$, we have the following equations:

$$\boldsymbol{\tau}_G = \boldsymbol{W}^T \boldsymbol{\tau}_O \tag{2.7}$$

$$\boldsymbol{\tau}_G = \boldsymbol{S}^T \boldsymbol{\tau}_A. \tag{2.8}$$

We can compute the actuator torque $\boldsymbol{\tau}_A$ from those of the tree structure $\boldsymbol{\tau}_O$ through the generalized force $\boldsymbol{\tau}_G$ once the sensitivity matrices $\boldsymbol{W}$ and $\boldsymbol{S}$ are computed, which is the subject of Section 2.4.

Nakamura et al. [63] did not use the generalized coordinates explicitly assuming that $\delta\boldsymbol{\theta}_G$ is taken as a subspace of $\delta\boldsymbol{\theta}_A$. As explained above, introducing the generalized coordinates eliminates unnecessary assumptions and restrictions on virtually cut joints, and on the placement of actuated joints.

2.4 Computation of $\boldsymbol{W}$ and $\boldsymbol{S}$

For many practical planar closed kinematic chains, $\boldsymbol{W}$ and $\boldsymbol{S}$ become constant and can be formed from visual inspection. It is also known that they are computed relatively easily for some special closed kinematic chains such as parallel mechanisms. In this subsection we provide a general method for computing the two matrices.

Consider a closed loop illustrated in Fig. 2.3. The linear and angular velocities of the shadowed link L is computed from $\dot{\boldsymbol{\theta}}_A$ as well as $\dot{\boldsymbol{\theta}}_B$ by multiplying the Jacobian

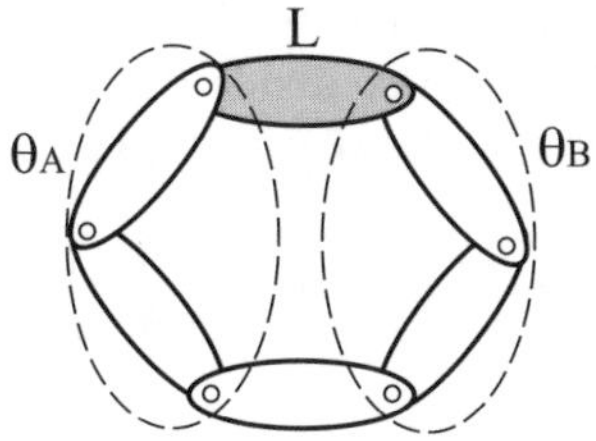

Fig. 2.3. A closed loop.

matrices $\boldsymbol{J}_A$ and $\boldsymbol{J}_B$ of the position and orientation of link L with respect to $\boldsymbol{\theta}_A$ and $\boldsymbol{\theta}_B$, respectively. The closed loop imposes the constraint that the velocity of link L computed from $\boldsymbol{\theta}_A$ should be equal to that from $\boldsymbol{\theta}_B$, namely,

$$\begin{pmatrix} \boldsymbol{J}_A & -\boldsymbol{J}_B \end{pmatrix} \begin{pmatrix} \dot{\boldsymbol{\theta}}_A \\ \dot{\boldsymbol{\theta}}_B \end{pmatrix} = \boldsymbol{O}. \tag{2.9}$$

Extending the discussion to the whole mechanism, the constraint due to the i-th closed loop is written in the form

$$\boldsymbol{J}_{Li}\dot{\boldsymbol{\theta}}_J = \boldsymbol{O} \tag{2.10}$$

where $\boldsymbol{J}_{Li}$ is a 6 by N_J matrix. The columns of $\boldsymbol{J}_{Li}$ consist of those of the Jacobian matrices of link L with respect to the joint angles, which can be calculated in the same way as serial kinematic chains [18].

Let N_L be the number of independent closed loop in the structure. Then we get N_L constraint matrices $\boldsymbol{J}_{Li}(i = 1, 2, \ldots, N_L)$, which forms the matrix $\boldsymbol{J}_C \in \boldsymbol{R}^{6N_L \times N_J}$ as

$$\boldsymbol{J}_C \triangleq \begin{pmatrix} \boldsymbol{J}_{L1} \\ \boldsymbol{J}_{L2} \\ \vdots \\ \boldsymbol{J}_{LN_L} \end{pmatrix}. \tag{2.11}$$

Although $\boldsymbol{J}_C$ represents all the kinematic constraints in the mechanism, not all the rows in $\boldsymbol{J}_C$ are independent, that is, $\boldsymbol{J}_C$ is not always full rank. We extract linearly independent rows from $\boldsymbol{J}_C$ and form $\boldsymbol{J}_{Cm} \in \boldsymbol{R}^{m \times N_J}$ where m is the rank of $\boldsymbol{J}_C$. From Eq.(2.10), $\boldsymbol{J}_{Cm}$ satisfies:

$$\boldsymbol{J}_{Cm}\dot{\boldsymbol{\theta}}_J = \boldsymbol{O}. \tag{2.12}$$

Equation (2.12) represents the m independent constraints of the closed loops. Since we have N_J joints under m constraints, the remaining degrees of freedom (mobility) of the whole mechanism N_F becomes

$$N_F = N_J - m. \tag{2.13}$$

Now we form $\boldsymbol{J}_S$ by extracting m independent columns from $\boldsymbol{J}_{Cm}$, and $\boldsymbol{J}_G$ by gathering the remaining columns. Also divide $\boldsymbol{\theta}_J$ into $\boldsymbol{\theta}_S$ and $\boldsymbol{\theta}_G$ according to the division of $\boldsymbol{J}_{Cm}$. From Eq.(2.12), $\boldsymbol{J}_S$, $\boldsymbol{J}_G$, $\boldsymbol{\theta}_S$ and $\boldsymbol{\theta}_G$ satisfy the equation

$$\boldsymbol{J}_{Cm}\dot{\boldsymbol{\theta}}_J = \begin{pmatrix} \boldsymbol{J}_S \ \boldsymbol{J}_G \end{pmatrix} \begin{pmatrix} \dot{\boldsymbol{\theta}}_S \\ \dot{\boldsymbol{\theta}}_G \end{pmatrix} = \boldsymbol{O}. \tag{2.14}$$

Equivalently,

$$\boldsymbol{J}_S\dot{\boldsymbol{\theta}}_S = -\boldsymbol{J}_G\dot{\boldsymbol{\theta}}_G. \tag{2.15}$$

Since $\boldsymbol{J}_S$ is always invertible, $\dot{\boldsymbol{\theta}}_S$ is uniquely determined by

$$\dot{\boldsymbol{\theta}}_S = \boldsymbol{H}\dot{\boldsymbol{\theta}}_G \tag{2.16}$$

$$\boldsymbol{H} \triangleq \frac{\partial \boldsymbol{\theta}_S}{\partial \boldsymbol{\theta}_G} = -\boldsymbol{J}_S^{-1}\boldsymbol{J}_G. \tag{2.17}$$

Equation (2.16) implies that we can choose $\boldsymbol{\theta}_G$ as the generalized coordinates. It is worth pointing out that the generalized coordinates are automatically selected through the process of forming $\boldsymbol{J}_S$. Note that $\boldsymbol{J}_S$, the independent columns of $\boldsymbol{J}_{Cm}$, corresponds to the *dependent* joint angles $\boldsymbol{\theta}_S$, not to the *independent* ones $\boldsymbol{\theta}_G$.

The Jacobian matrices $\boldsymbol{W}$ and $\boldsymbol{S}$ are formed from $\boldsymbol{H}$ immediately as follows:

- $\boldsymbol{W}$: If the ith joint of $\boldsymbol{\theta}_O$ is not a member of the generalized coordinates and corresponds with the jth one of $\boldsymbol{\theta}_S$, then include the jth row of $\boldsymbol{H}$ as the ith row of $\boldsymbol{W}$. If it is a member of the generalized coordinates and corresponds with the jth joint of $\boldsymbol{\theta}_G$, then include a unit vector with jth element being 1 and others 0 as the ith row of $\boldsymbol{W}$. This procedure is shown in Fig. 2.4.
- $\boldsymbol{S}$: If the ith joint of $\boldsymbol{\theta}_A$ is not a member of the generalized coordinates and corresponds with the jth one of $\boldsymbol{\theta}_S$, then include the jth row of $\boldsymbol{H}$ as the ith row of $\boldsymbol{S}$. If it is a member of the generalized coordinates and corresponds with the jth joint of $\boldsymbol{\theta}_G$, then include a unit vector with jth element being 1 and others 0 as the ith row of $\boldsymbol{S}$. This procedure is shown in Fig. 2.5

2.5 Relationship of Accelerations

Differentiating Eq.(2.16) by time yields

$$\ddot{\boldsymbol{\theta}}_S = \boldsymbol{H}\ddot{\boldsymbol{\theta}}_G + \dot{\boldsymbol{H}}\dot{\boldsymbol{\theta}}_G \tag{2.18}$$

which calculates the acceleration of dependent joints $\ddot{\boldsymbol{\theta}}_S$ from the generalized acceleration $\ddot{\boldsymbol{\theta}}_G$. This computation is required in forward dynamics computation. In this subsection, computation of the second term of the right-hand side of Eq.(2.18) is presented.

From Eq.(2.17) we have

$$\dot{\boldsymbol{H}}\dot{\boldsymbol{\theta}}_G = -\left\{\frac{d}{dt}(\boldsymbol{J}_S^{-1})\boldsymbol{J}_G + \boldsymbol{J}_S^{-1}\dot{\boldsymbol{J}}_G\right\}\dot{\boldsymbol{\theta}}_G. \tag{2.19}$$

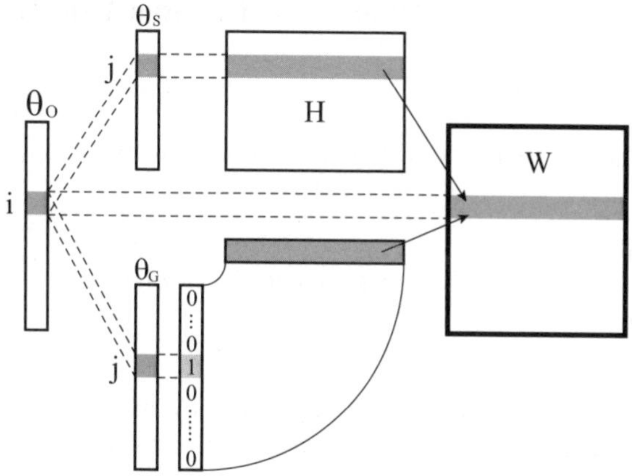

Fig. 2.4. Forming $\boldsymbol{W}$ from $\boldsymbol{H}$.

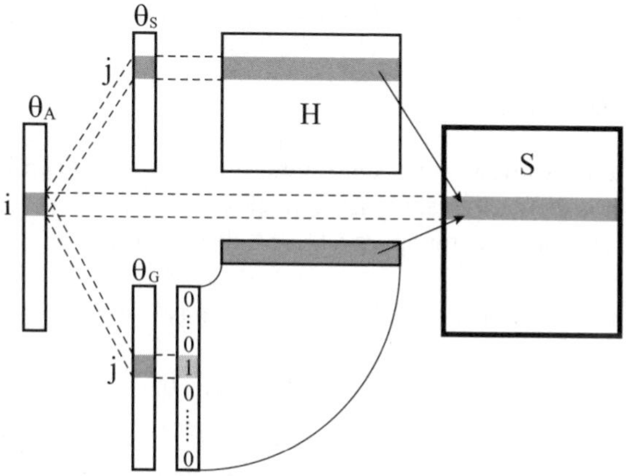

Fig. 2.5. Forming $\boldsymbol{S}$ from $\boldsymbol{H}$

On the other hand, $\boldsymbol{J}_S^{-1}\boldsymbol{J}_S = \boldsymbol{E}$, where $\boldsymbol{E}$ represents the identity matrix, yields

$$\frac{d}{dt}(\boldsymbol{J}_S^{-1})\boldsymbol{J}_S + \boldsymbol{J}_S^{-1}\dot{\boldsymbol{J}}_S = \boldsymbol{O}. \tag{2.20}$$

Using Eqs.(2.15) and (2.20), Eq.(2.19) becomes

$$\begin{aligned}\dot{\boldsymbol{H}}\dot{\boldsymbol{\theta}}_G &= -\boldsymbol{J}_S^{-1}(\dot{\boldsymbol{J}}_S\dot{\boldsymbol{\theta}}_S + \dot{\boldsymbol{J}}_G\dot{\boldsymbol{\theta}}_G) \\ &= -\boldsymbol{J}_S^{-1}\dot{\boldsymbol{J}}_{Cm}\dot{\boldsymbol{\theta}}_J. \end{aligned} \tag{2.21}$$

$\dot{\boldsymbol{J}}_{Cm}\dot{\boldsymbol{\theta}}_J$ is formed by extracting the elements of $\dot{\boldsymbol{J}}_C\dot{\boldsymbol{\theta}}_J$ corresponding to $\boldsymbol{J}_{Cm}$, where $\dot{\boldsymbol{J}}_C\dot{\boldsymbol{\theta}}_J$ is computed in the same algorithm as one for serial manipulators [69].

2.6 Inverse and Forward Dynamics of Closed Kinematic Chains

2.6.1 Inverse Dynamics

The inverse dynamics computation of general closed kinematic chains consists of the following four steps:

(1) Compute $\boldsymbol{W}$ and $\boldsymbol{S}$
(2) Compute $\boldsymbol{\tau}_O$ by inverse dynamics computation for the tree structure
(3) Compute $\boldsymbol{\tau}_G$ by Eq.(2.7)
(4) Compute $\boldsymbol{\tau}_A$ by solving the linear equation (2.8)

If the mechanism does not have actuation redundancy, namely, if the number of actuators equals to the degrees of freedom, $\boldsymbol{S}$ becomes a square matrix. Thus, $\boldsymbol{\tau}_A$ is computed by

$$\boldsymbol{\tau}_A = \boldsymbol{S}^{-T}\boldsymbol{W}^T\boldsymbol{\tau}_O. \tag{2.22}$$

Otherwise $\boldsymbol{\tau}_A$ is not determined uniquely, and some optimization method should be applied. Refer to [67] for methods of optimizing actuation redundancy.

2.6.2 Forward Dynamics

Although several forward dynamics algorithms are known for open kinematic chains [31, 1, 22, 96], it is difficult to apply them to closed chains due to the complexity of their structure. The unit vector approach [96], however, can be extended to closed chains easily.

The equation of motion of closed kinematic chains is written in the same form as open chains as

$$\boldsymbol{\tau}_G = \boldsymbol{A}(\boldsymbol{\theta}_G)\ddot{\boldsymbol{\theta}}_G + \boldsymbol{b}(\boldsymbol{\theta}_G, \dot{\boldsymbol{\theta}}_G) \tag{2.23}$$

where $\boldsymbol{\tau}_G \in \boldsymbol{R}^{N_F}$ is the generalized force, $\boldsymbol{A} \in \boldsymbol{R}^{N_F \times N_F}$ is the inertia matrix and $\boldsymbol{b} \in \boldsymbol{R}^{N_F}$ includes the sum of centrifugal, Coriolis and gravitational forces. In open kinematic chains, the joint angles are usually used as the generalized coordinate and thus the joint torques are the generalized force. Therefore, the accelerations of all joints are computed directly by Eq.(2.23). In closed kinematic chains, on the other hand, the joint torque vector and the generalized force may differ even in their dimensions. Additional computations of transformation of the joint torques into the generalized force and the generalized acceleration into the joint acceleration are required.

The forward dynamics algorithm based on the inverse dynamics algorithm explained in Section 2.3 and the unit vector approach is summarized as follows:

(1) Transform the input joint torques $\boldsymbol{\tau}_A$ into the generalized force $\boldsymbol{\tau}_G$ by Eq.(2.8).
(2) Compute the inverse dynamics for the zero generalized acceleration and let the resultant generalized force be $\boldsymbol{b}$. Using Eq.(2.18), the accelerations of the dependent joints $\ddot{\boldsymbol{\theta}}_S$ are given by $\dot{\boldsymbol{H}}\dot{\boldsymbol{\theta}}_G$, whose computation method is shown in section 2.5.

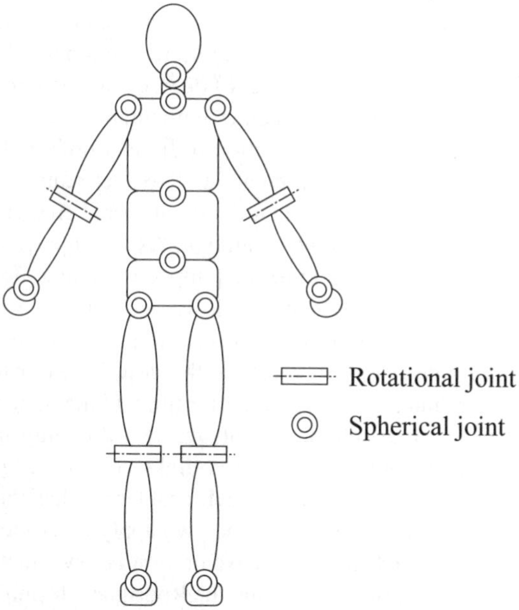

Fig. 2.6. An example of human model with rotational and spherical joints.

(3) Execute the following computation for $i = 1, 2, \ldots, N_F$:
  a) Compute the inverse dynamics with $\ddot{\boldsymbol{\theta}}_G = \boldsymbol{e}_i$, where $\boldsymbol{e}_i \in \boldsymbol{R}^{N_F}$ is a unit vector whose i-th element is 1 and others 0. The accelerations of the dependent joints are computed by substituting $\boldsymbol{e}_i$ for $\ddot{\boldsymbol{\theta}}_G$ in Eq.(2.18).
  b) Let the computed generalized force be $\boldsymbol{f}_i$ and calculate $\boldsymbol{a}_i$ by $\boldsymbol{a}_i = \boldsymbol{f}_i - \boldsymbol{b}$. We can obtain $\boldsymbol{a}_i$ directly by computing the inverse dynamics with $\ddot{\boldsymbol{\theta}}_G = \boldsymbol{e}_i, \dot{\boldsymbol{\theta}}_G = 0$ and no gravity, which would save the computational cost a little.
  c) Include $\boldsymbol{a}_i$ as the i th column of $\boldsymbol{A}$.

(4) Using $\boldsymbol{\tau}_G, \boldsymbol{b}$ and $\boldsymbol{A}$, compute the generalized acceleration by

$$\ddot{\boldsymbol{\theta}}_G = \boldsymbol{A}^{-1}(\boldsymbol{\tau}_G - \boldsymbol{b}). \tag{2.24}$$

(5) Compute $\ddot{\boldsymbol{\theta}}_S$ by Eq.(2.18), where $\dot{\boldsymbol{H}}\dot{\boldsymbol{\theta}}_G$ is already computed in step 2, to get the accelerations of all joints.

2.7 Multi-degrees-of-Freedom Joints

2.7.1 Spherical Joints

Fig. 2.6 is a typical human body model composed of mechanical joints. The 40 degrees of freedom of the model include 4 rotational joints and 12 spherical joints, which shows that many joints in human bodies can be modeled as spherical joints.

In robot manipulators, a spherical joint is mechanically implemented as three successive rotational joints with their axes intersecting at a point. With this mechanical implementation and the modeling, the 40 degrees of freedom model of human body would need 41 links to treat in dynamics computation.

Physiological structure or implementation of human body is far more complex and beyond our scope of efficient computation. This fact requires and allows us to adopt a mechanical model that is suitable for computational efficiency and not necessarily constrained by mechanical implementation. As a computational model of human figures, we assume a spherical joint is equipped with three-degrees-of-freedom spherical motor or a similar actuator. With this assumption, we can significantly reduce the number of links. In fact, only 17 links are required for the model in Fig. 2.6 if spherical joints are considered. In addition, the description of link configuration becomes simpler and requires no discussion of artificial kinematic singularity.

Three-degrees-of-freedom spherical joints cause a difficulty in numerical integration of relative orientation between the two links connected by them. Although the Euler angles representation can avoid such problem, it has the problem of singularity. Integration problem would arise when we apply other methods such as the Euler parameters [18] to avoid singularity. We present below a method of first-order Euler integration of relative orientation using the Rodrigues' formulation [18], which is often used for finite spatial rotation.

Let $\boldsymbol{\omega}_i$ be the relative angular velocity and $\boldsymbol{R}_i$ the relative orientation of link i with respect to its parent link at time t. The relative orientation at $t + \Delta t$, $\boldsymbol{R}_i'$, is computed by

$$\boldsymbol{R}_i' = (\boldsymbol{E}_3 + \boldsymbol{\Omega} \sin\theta + \boldsymbol{\Omega}^2(1 - \cos\theta))\boldsymbol{R}_i \tag{2.25}$$

where

$$\boldsymbol{\theta} = \boldsymbol{\omega}_i \Delta t \tag{2.26}$$

$$\theta = |\boldsymbol{\theta}| \tag{2.27}$$

$$\theta(\bar{\omega}_x\, \bar{\omega}_y\, \bar{\omega}_z)^T = \boldsymbol{\theta} \tag{2.28}$$

$$\text{if } \theta \neq 0, \quad \boldsymbol{\Omega} = \begin{pmatrix} 0 & -\bar{\omega}_z & \bar{\omega}_y \\ \bar{\omega}_z & 0 & -\bar{\omega}_x \\ -\bar{\omega}_y & \bar{\omega}_x & 0 \end{pmatrix} \tag{2.29}$$

$$\text{if } \theta = 0, \quad \boldsymbol{\Omega} = \boldsymbol{O} \tag{2.30}$$

and $\boldsymbol{E}_3$ is a 3×3 identity matrix.

Spherical joints may be regarded as a combination of three rotational joints whose axes lies on the x, y and z axes of the link coordinate. Therefore relative angular velocities and accelerations of the two links expressed in link coordinate corresponds to the joint velocity and acceleration of a single-degree-of-freedom joint.

2.7.2 Free Joints

In order to treat the cases where the base link is not fixed to the inertial frame, we introduce a six-degrees-of-freedom "free" joint between the base link and the inertial

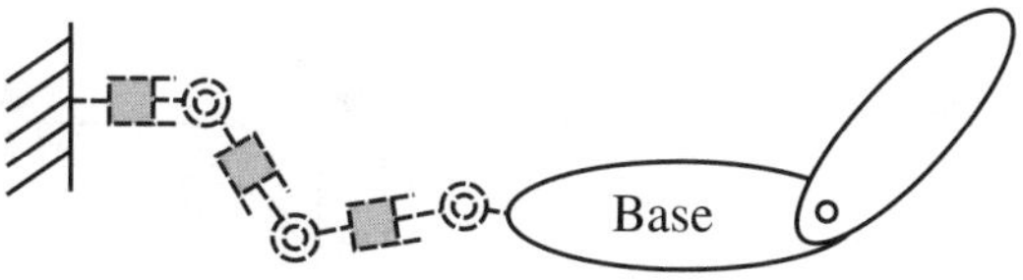

Fig. 2.7. A free-flying kinematic chain.

frame, whose actuator torque is always zero; thus forward dynamics is computed in the same way as base-fixed chains. Note that the six-degrees-of-freedom joint is not decomposed into six single-degree-of-freedom joints but treated as one joint. This can reduce the amount of computations especially for the recursive computation of kinematics and dynamics.

Free joints may be regarded as a combination of three linear joints that can move in x, y and z directions, and a spherical joint. Therefore the linear and angular velocities and accelerations of the free link expressed in link coordinate corresponds to the joint velocity and acceleration of a single-degree-of-freedom joint. Integrating the angular elements of joint velocities and accelerations is done in just the same way as spherical joints. The three elements of the linear part can be integrated independently as in three separate linear joints.

2.8 Examples

2.8.1 Computation of W and S

We first show an example of computing the matrices $\boldsymbol{W}$ and $\boldsymbol{S}$ for a simple structure. Consider the four-bar linkage in Fig. 2.8(a) with four parallel rotational joints 1–4. Suppose joints 1 and 3 are actuated and joints 2 and 4 are not. For this simple mechanism, it is obvious from a geometric intuition that the structure is planar and has 1DOF (so redundantly actuated). The Jacobian matrices $\boldsymbol{W}$ and $\boldsymbol{S}$ are constant for any configuration and can be easily obtained. However, we start without any assumption and treat it as a general spatial mechanism, and show that we can obtain the same result by simply following our method.

First we derive the constraint equation Eq.(2.12) for the closed loop. Using the condition that the velocities of *Link B* computed from $(\dot{\theta}_1\dot{\theta}_2)$ and $(\dot{\theta}_3\dot{\theta}_4)$ should be the same, we have

$$\begin{pmatrix} 0 & 0 \\ 1 & 0 \\ 1 & 1 \\ 1 & 1 \\ 0 & 0 \\ 0 & 0 \end{pmatrix} \begin{pmatrix} \dot{\theta}_1 \\ \dot{\theta}_2 \end{pmatrix} = \begin{pmatrix} 0 & 0 \\ 1 & 0 \\ 0 & 0 \\ 1 & 1 \\ 0 & 0 \\ 0 & 0 \end{pmatrix} \begin{pmatrix} \dot{\theta}_3 \\ \dot{\theta}_4 \end{pmatrix} \tag{2.31}$$

which is rewritten as

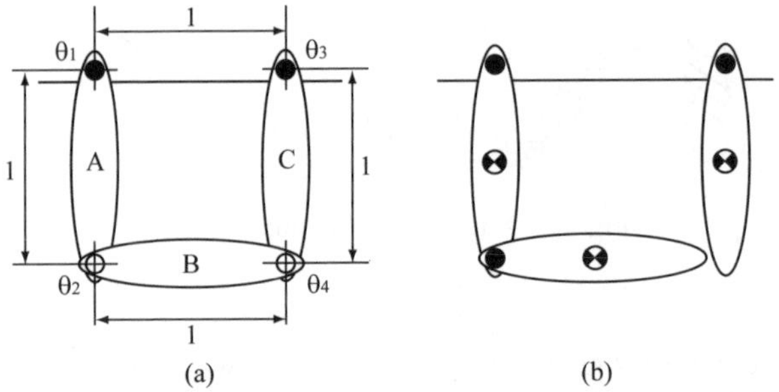

Fig. 2.8. A four-bar closed kinematic chain (a) and its associated virtual open kinematic chain (b).

$$\begin{pmatrix} 0 & 0 & 0 & 0 \\ 1 & 0 & -1 & 0 \\ 1 & 1 & 0 & 0 \\ 1 & 1 & -1 & -1 \\ 0 & 0 & 0 & 0 \\ 0 & 0 & 0 & 0 \end{pmatrix} \begin{pmatrix} \dot{\theta}_1 \\ \dot{\theta}_2 \\ \dot{\theta}_3 \\ \dot{\theta}_4 \end{pmatrix} = \boldsymbol{O}. \tag{2.32}$$

The rank of the coefficient matrix of Eq.(2.32) obviously three. From this fact, we have confirmed the intuition that this structure has $4 - 3 = 1$DOF. We can extract the three independent rows and obtain

$$\begin{pmatrix} 1 & 0 & -1 & 0 \\ 1 & 1 & 0 & 0 \\ 1 & 1 & -1 & -1 \end{pmatrix} \begin{pmatrix} \dot{\theta}_1 \\ \dot{\theta}_2 \\ \dot{\theta}_3 \\ \dot{\theta}_4 \end{pmatrix} = \boldsymbol{O}. \tag{2.33}$$

Next we split the joint angles $(\dot{\theta}_1\ \dot{\theta}_2\ \dot{\theta}_3\ \dot{\theta}_4)^T$ into independent and dependent coordinates by selecting three independent columns of the coefficient matrix of Eq.(2.33). Any selection of three columns works here, but let us choose 1st, 2nd and 4th columns:

$$\begin{pmatrix} 1 & 0 & 0 \\ 1 & 1 & 0 \\ 1 & 1 & -1 \end{pmatrix} \begin{pmatrix} \dot{\theta}_1 \\ \dot{\theta}_2 \\ \dot{\theta}_4 \end{pmatrix} = \begin{pmatrix} 1 \\ 0 \\ 1 \end{pmatrix} \dot{\theta}_3. \tag{2.34}$$

Solving Eq.(2.34) for $(\dot{\theta}_1\ \dot{\theta}_2\ \dot{\theta}_4)^T$ yields

$$\begin{pmatrix} \dot{\theta}_1 \\ \dot{\theta}_2 \\ \dot{\theta}_4 \end{pmatrix} = \begin{pmatrix} 1 & 0 & 0 \\ 1 & 1 & 0 \\ 1 & 1 & -1 \end{pmatrix}^{-1} \begin{pmatrix} 1 \\ 0 \\ 1 \end{pmatrix} \dot{\theta}_3 \tag{2.35}$$

$$= \begin{pmatrix} 1 & 0 & 0 \\ -1 & 1 & 0 \\ 0 & 1 & -1 \end{pmatrix}^{-1} \begin{pmatrix} 1 \\ 0 \\ 1 \end{pmatrix} \dot{\theta}_3 \tag{2.36}$$

$$= \begin{pmatrix} 1 \\ -1 \\ -1 \end{pmatrix} \dot{\theta}_3. \tag{2.37}$$

This result also coincides with the intuition because, when θ_3 increases, θ_1 increases accordingly while θ_2 and θ_4 decreases with the same speed.

Because joints 1 and 3 are actuated, $\boldsymbol{S}$ is formed as

$$\boldsymbol{S} = \begin{pmatrix} 1 \\ 1 \end{pmatrix}. \tag{2.38}$$

If we cut joint 4 for the inverse dynamics computation, $\boldsymbol{W}$ is formed as

$$\boldsymbol{W} = \begin{pmatrix} 1 \\ -1 \\ 1 \end{pmatrix}. \tag{2.39}$$

The inverse dynamics computation is performed as follows. Suppose we need the joint torques to keep the mechanism in the configuration shown in Fig. 2.8(a). Assuming the mass of each link is 1 with the center of mass at the middle of the link, the inverse dynamics of the two open kinematic chains shown in Fig. 2.8(b) would result in

$$\begin{pmatrix} \hat{\tau}_1 \\ \hat{\tau}_2 \\ \hat{\tau}_4 \end{pmatrix} = \begin{pmatrix} 4.9 \\ 4.9 \\ 0 \end{pmatrix} \tag{2.40}$$

where $\hat{\tau}_1, \hat{\tau}_2$ and $\hat{\tau}_4$ denotes the joint torques at joints 1, 2 and 4, respectively, for the virtual open kinematic chain. These torques are converted to the generalized force using $\boldsymbol{W}$ as

$$\tau_G = \boldsymbol{W}^T \begin{pmatrix} \hat{\tau}_1 \\ \hat{\tau}_2 \\ \hat{\tau}_4 \end{pmatrix} = \begin{pmatrix} 1 & -1 & 1 \end{pmatrix} \begin{pmatrix} 4.9 \\ 4.9 \\ 0 \end{pmatrix} = 0. \tag{2.41}$$

The relationship between the generalized force τ_G and the actuator torques τ_1 and τ_3 is written as

$$\tau_G = \boldsymbol{S}^T \begin{pmatrix} \tau_1 \\ \tau_3 \end{pmatrix} = \tau_1 + \tau_3 \tag{2.42}$$

Eqs.(2.41) and (2.42) indicates that the sum of the actuator torques should be zero to keep current configuration, which again coincides with the intuition.

2.8.2 Simulation Example

Fig. 2.9 shows an example of dynamics computation of closed kinematic chains. The human figure (monkey) has 16DOF and the swing consists of 12DOF with one closed loop.

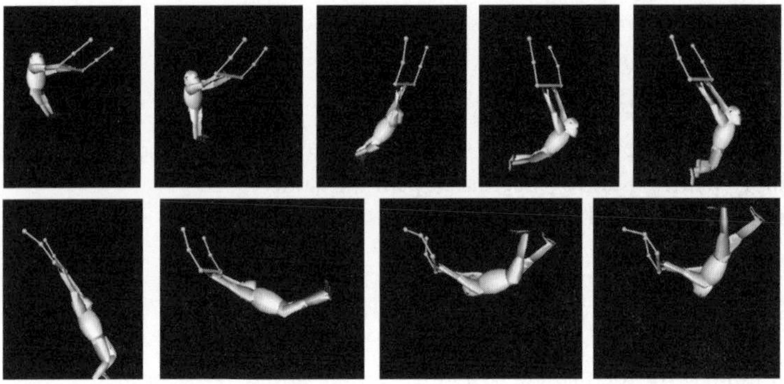

Fig. 2.9. An example of dynamics simulation of closed kinematic chains.

2.9 Summary

This chapter presented the methods for inverse and forward dynamics computation of general closed kinematic chains. The contributions of this chapter are summarized as follows:

(1) The concept of the generalized coordinate of closed kinematic chains was proposed. Introducing generalized coordinates eliminates unnecessary assumptions on the virtually cut joints and actuated joints.
(2) General algorithms for computing the Jacobian matrices for force transformation were presented. The method also computes the degrees of freedom and selects the generalized coordinates at the same time.
(3) Inverse and forward dynamics algorithm was formulated for general closed kinematic chains. The inertial matrix in joint space, used for forward dynamics computation, is computed by applying unit vector method [96].
(4) An approach was proposed for treating multi-DOF joints as a single joint to reduce the computation. Especially, scheme for integrating angular velocity using Rodrigues' formulation was proposed.

The techniques presented in this chapter are to be applied to structure-varying kinematic chains in Chapter 3, and then to collision/contact simulation in Chapter 5. The total simulation method is also applied to dynamics filter as described in Chapter 7.

The problem of this method is that it requires $O(N^3)$ computations, where N is the degrees of freedom of the system. The method presented in Chapter 4 improves the complexity to $O(N)$ for serial computation and $O(\log N)$ for parallel computation on $O(N)$ processors.

3

Link Connectivity Desription for Structure-Varying Kinematic Chains

3.1 Introduction

This chapter presents a method for describing link connectivity of structure-varying kinematic chains. Link connectivity description scheme consists of the following two issues:

(1) how to describe open and closed kinematic chain in a consistent way, and
(2) how to update the link connectivity state in accordance with the structural changes.

Pointers and *virtual links* are used for describing open and closed kinematic chains. A pointer acts as an arrow from a link to another in a specific relationship. A virtual link represents a closed loop and indicates where the loop is virtually cut for dynamics computation described in Chapter 2. This description scheme has an advantage over some methods in terms of required memory. Matrix representation, for example, where each element of the matrix indicates whether a pair of links are connected, requires $O(N^2)$ data where N denotes the number of links, while method using pointers requires $O(N)$ data because each link has no more than 4 pointers including one to indicate the real link of a virtual link.

Structural changes, both connecting two links and cutting a joint, are handled in an uniform way. All link connections and joint cuts are processed by adding and removing a virtual link, respectively. A link connection, therefore, always forms a closed kinematic chain even if it appears to be an open one. This scheme leads to the simple maintenance of link connectivity description by avoiding inversion of parent-child relationships of links, thus the overhead due to the modification of connectivity data is kept small.

When two links are connected with nonzero relative velocity, the impulsive force due to collision causes the discontinuous changes of the joint velocities to satisfy

This chapter was adapted from, by permission, Y. Nakamura and K. Yamane, "Dynamics Computation of Structure-Varying Kinematic Chains and Its Application to Human Figures," IEEE Transactions on Robotics and Automation, vol.16, no.2, pp.124–134, 2000.

the constraint of the new joint. The new joint velocities are then used as the velocity boundary condition for the next step. We present a method for computing the velocity boundary condition after a collision based on the equations of kinematic constraint and conservation of momentum.

This chapter is organized as follows. First, Section 3.3.1 presents the method to describe open kinematic chains using pointers, which is then extended to closed kinematic chains by introducing virtual links in Section 3.3.2. Section 3.3.3 discusses how to update the connectivity description when the real structural changes. Section 3.4 presents the method for computing the velocity boundary condition after link connections with nonzero relative velocity. Finally, the link connectivity description is combined with the forward dynamics algorithm presented in the previous chapter and several simulation examples demonstrate the effectiveness of the scheme.

3.2 Related Work

Seamless dynamics computation of structure-varying kinematic chains would be realized by a combination of a dynamics computation algorithm applicable to general kinematic chains without any modification of the program, and a connectivity description that enables automatic and efficient modification of the internal connectivity data. Not all the dynamics computation algorithms are qualified for the application to structure-varying kinematic chains. Symbolic methods [34], for example, are inherently impossible to handle general structure-varying kinematic chains because they have to switch among different programs when the link connectivity changes.

Although link connectivity description is not emphasized very much in literature, it is closely related to the efficiency of implementation and, as revealed later, seamless handling of structural changes. Existing description schemes include linear graphs [58], matrices [1], and vectors [22]. They are indeed suitable for describing an algorithm systematically on paper. From a practical or programming point of view, however, they are not always effective because the program has to search among the elements to find out whether there exists a closed loop or even which link is connected to a specific link.

Dynamics computation of general structure-varying kinematic chains has not been discussed in literature to the author's knowledge. Locomotion of multilegged robots, however, may be regarded as a subset of structure-varying kinematic chains. Surla et al. [93] combined symbolic equations of motions for all phases in walking. Perrin et al. [76] applied two basic algorithms to simulation of biped and quadruped robots. Mcmillan et al. [57] used a spring-damper model to compute the constraint forces between the legs and the ground.

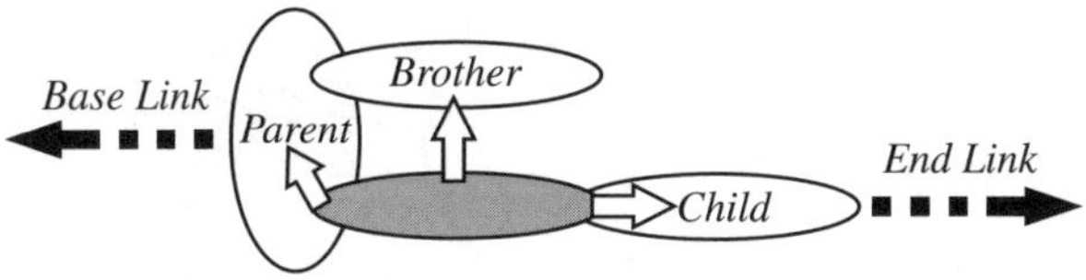

Fig. 3.1. Three pointers to describe open kinematic chains.

```
void forward_path(Link target) {
    if(target == NULL) return;          // terminate if this is an empty link
    forward_path_main();                // main computation
    forward_path(target.child);         // recursive call to the child link
    forward_path(target.brother);       // recursive call to the brother link
}

void backward_path(Link target) {
    if(target == NULL) return;          // terminate if this is an empty link
    backward_path(target.parent);       // recursive call to the parent link
    backward_path_main();               // main computation
}
```

Fig. 3.2. A pseudocode for implementing forward- and backward-path computations using pointers.

3.3 Link Connectivity Description and Its Maintenance

3.3.1 Pointers Describe Open Kinematic Chains

For the efficiency of computation, and for the convenience of implementation, we propose to use *pointers* to describe the link connectivity. *Pointer* is an important function of C/C++ programming language and acts as an arrow from a link to another. Because the actual value of a pointer is the address of a specified datum, we can refer to the data of a link immediately through the pointer to its data.

We use three pointers for each link to describe open kinematic chains. The meanings of the pointers are illustrated in Fig. 3.1: the *parent* pointer points the next link connected towards the base link, the *child* pointer points the next link connected towards an end link, and the *brother* pointer points a link with the same *parent* in case the parent link has several links connected towards end links.

The recursive dynamics computations of the Newton-Euler formulation [52] are implemented using the three pointers and recursive call of functions. For the forward path computations, the functions are called recursively for the *child* and *brother* links after the computation for itself. For the backward path computations, on the other hand, recursive calls are made before the computation for itself. The pseudocode in Fig. 3.2 illustrates how to implement forward- and backward-path computations using the pointers.

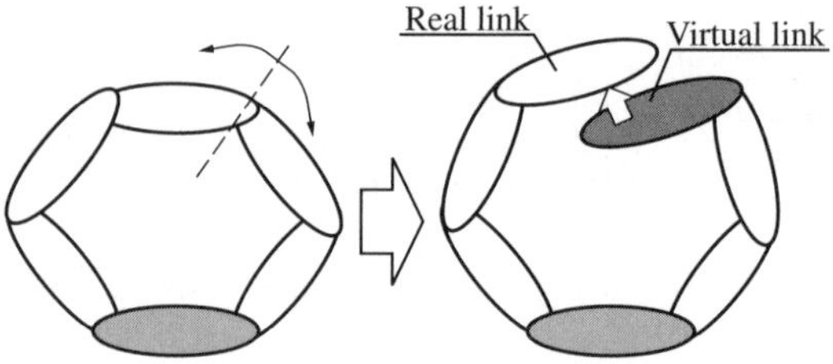

Fig. 3.3. Describing a closed loop by a virtual link.

3.3.2 Virtual Links Describe Closed Kinematic Chains

The three pointers are applicable only to open kinematic chains because the parent-child relationship for a closed kinematic chain results in an infinite loop.

First, as illustrated in Fig. 3.3, we virtually cut a joint in each closed loop to avoid infinite loops. We can describe the resulting mechanism by the three pointers because it is no longer a closed chain. To represent the connection at the virtually cut joints, we add a *virtual* link to one of the two links that had been connected by the cut joint. A virtual link has kinematic properties such as joint values and link length, but no dynamic properties such as mass or inertia. In order to indicate the real link of a virtual link, we introduce a new pointer called *real* pointer. The *real* pointer is valid only for virtual links. Note that a closed kinematic chain may be described in different ways depending on which joint in a closed loop is virtually cut.

Each virtual link represents the kinematic constraint of a closed loop that a virtual link and its corresponding real link should be at the same position and orientation. When the dynamics computation part finds a virtual link, it recognizes the real link through the *real* pointer.

To summarize, any open or closed kinematic chains are described by four pointers — *parent, child, brother* and *real* — and a *virtual* link corresponding to each closed loop. An example of a description of a closed kinematic chain is shown in Fig. 3.4. Our scheme has the following advantages:

- Suited for recursive implementation of dynamics computations.
- Easy to find closed loops because each closed loop has a corresponding virtual link.
- Simple choice of virtually cut joints for dynamics computation. They coincide with the joints of the virtual links.
- Less data and computation for link connectivity. They are proportional to the number of links.

3.3.3 Link Connectivity Maintenance

First, consider a case in which two links are connected to create a new joint. If a closed loop is generated by the connection, as in a case illustrated in Fig. 3.5, we add a virtual link at the new joint. The procedure is as simple as follows:

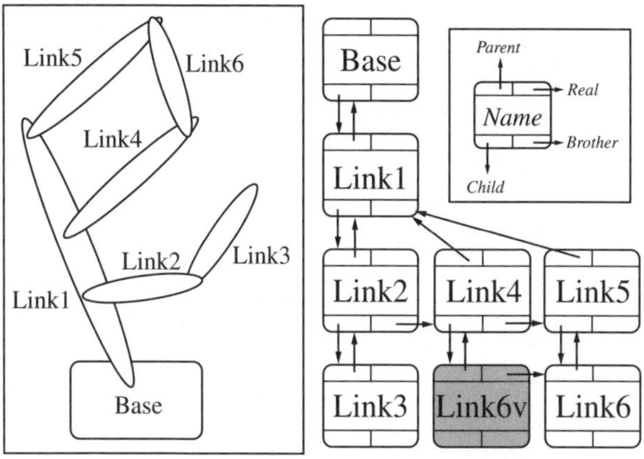

Fig. 3.4. A description of a closed kinematic chain.

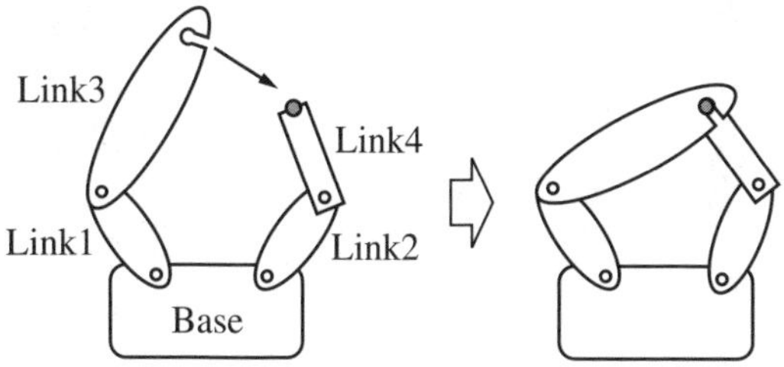

Fig. 3.5. Connection of two links.

(1) Create virtual link *Link 4v* whose real link is *Link 4*.
(2) Add *Link 4v* as a child of *Link 3*.

which is easily programmed and computed on line. The descriptions of link configurations before and after the connection are shown in Fig. 3.6.

In the case where a free-floating chain is connected to another chain, the situation becomes complicated. Fig. 3.7 shows a case where *Link 1* of a free-flying chain is connected to *Ground* and a new joint is created. Since the structure after the connection is apparently an open chain, it seems natural to change the data as shown in Fig. 3.8. The remarks "*Rotate*" and "*Free*" in the figure indicate the joint types. One must notice, however, that it requires the inversion of the parent-child relationship of *Base* and *Link 1*, which requires modification of the Denavit-Hartenberg parameters, the values of some dynamic parameters, and the indexing of joints. The modification is not difficult but needs additional computation, which is crucial if the structure varies in real time. When the structural change is known beforehand, the computational burden is reduced by preparing different connectivity models in advance, which would be, however, as tedious and complicated as switching between different dynamical models and algorithms.

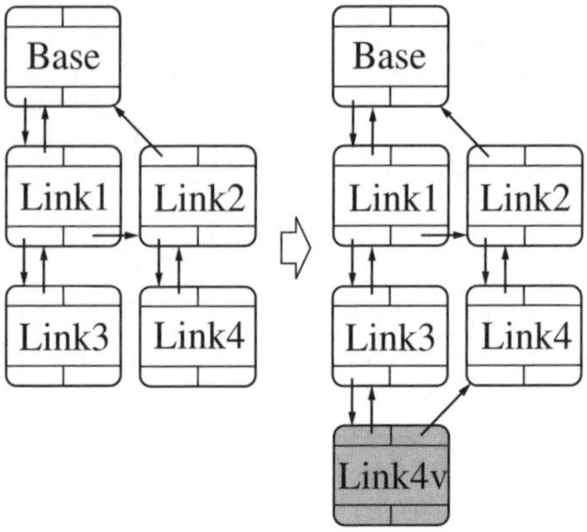

Fig. 3.6. Descriptions of the link connectivity before and after connection of Fig. 3.5.

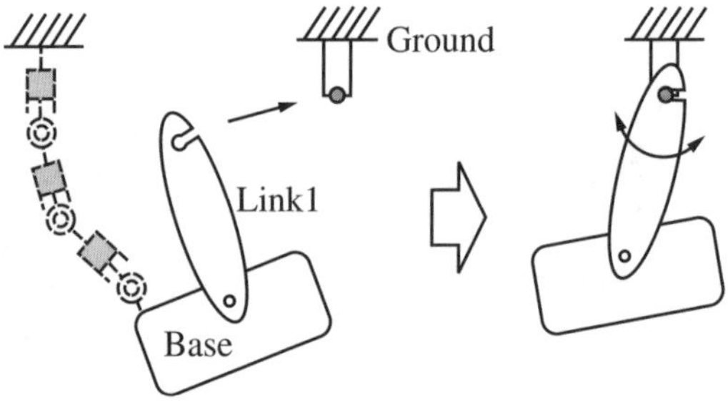

Fig. 3.7. An open kinematic chain generated by a link connection.

We propose to treat this case exactly in the same way as the previous case:

(1) Create a virtual link of *Link 1* and name it *Link 1v*.
(2) Connect *Link 1v* to *Ground* through the new rotational joint.

Fig. 3.9 shows the description of new structure, which does not require the inversion of the relationship of *Base* and *Link 1*.

Note that the number of joints increases only by one by handling the free joint as a single 6 DOF joint as explained in Section 2.7.2, although the amount of dynamics computation in this case becomes larger than when it is treated as an open chain. Therefore, we might need more careful comparison of computational loss due to easy connectivity maintenance and computational gain due to increase of the number of joints. However, we claim the advantage of the above procedure from the following two viewpoints:

(1) Simplicity of algorithm is valuable for programming and, eventually, offers better usability for the end-users.

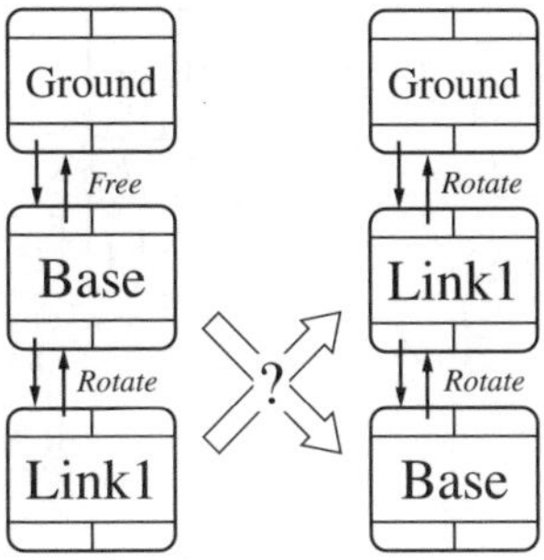

Fig. 3.8. A possible modification of link connectivity description for the link connection in Fig. 3.7.

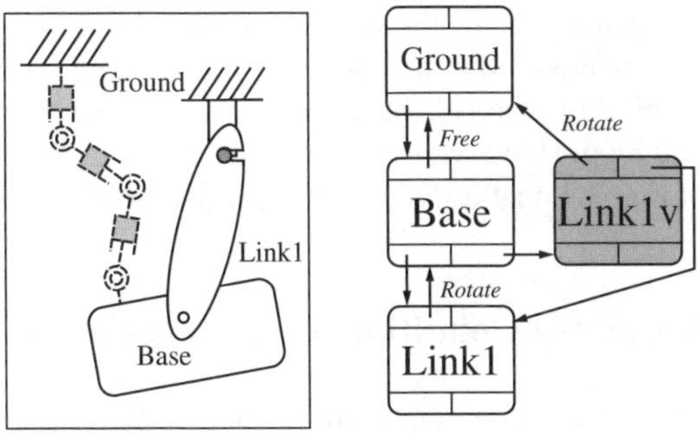

Fig. 3.9. A closed kinematic chain with a free joint.

(2) The computational gain due to increase of the number of joints would be reduced in time by employing parallel processing, while the computation for connectivity maintenance cannot take the advantage of parallelism.

In the rest half of this subsection we discuss the procedure for cutting a connection of two links at the joint between them.

If the cut joint was connecting a virtual link and its parent, the procedure is exactly the opposite of that in link connection. In the structure after the connection in Fig. 3.5, for example, cutting the joint between *Link 3* and *Link 4* can be handled simply by deleting *Link 4v*. We can generally assume that cutting occurs only at the joints connecting virtual links as long as human figures are concerned.

In general kinematic chains, however, this is not always the case. Even if the cut joint is not related to a virtual link, structural change can be easily handled by introducing a free joint as in Section 2.7.2. Suppose, in the structure after the connection in Fig. 3.5, that the joint between *Link 1* and *Link 3* is cut. The procedure in this case becomes:

(1) Cut the parent-child relation between *Link 1* and *Link 3*.
(2) Connect *Link 3* to *Base* by a free joint.

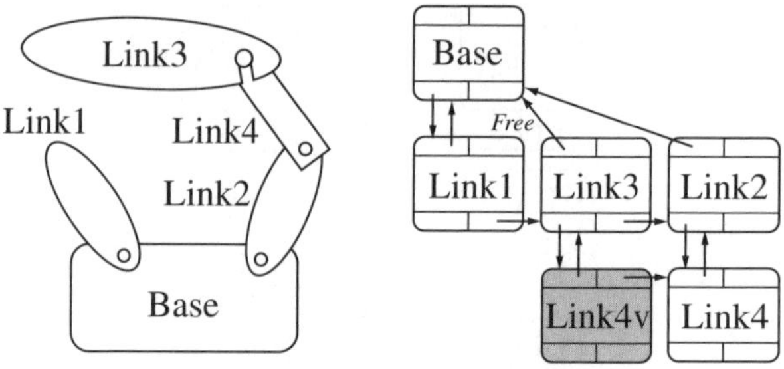

Fig. 3.10. The link structure and its description after a joint cut.

The link structure and its description are shown in Fig. 3.10. The connection between *Link 3* and *Link 4* is maintained by the virtual link *Link 4v*.

An alternative to deal with this situation would be to cut *Link 3* from *Link 1* and set as a child of *Link 4* in place of the virtual link *Link 4v*, in which case we can eliminate the closed loop. However, this scheme has the same problem as discussed in the examples of Fig. 3.7, that is, the inversion of parent-child relationship.

3.4 Velocity Boundary Condition after Structural Changes

When two links are connected with nonzero relative velocity, discontinuous change of joint velocities occurs due to the collision. The forward dynamics computation requires the boundary condition of joint velocities after the structural changes. In this section, we present an algorithm to compute the velocity boundary condition.

Suppose that the two connecting links belong to chain 1 and chain 2. Let $\boldsymbol{\theta}_i$ $(i = 1, 2)$ be the generalized coordinates of chains 1 and 2, $\boldsymbol{J}_i = \partial \boldsymbol{r}/\partial \boldsymbol{\theta}_i$ the Jacobian matrices of the connection point $\boldsymbol{r}$ with respect to the generalized coordinates, and $\boldsymbol{A}_i$ their inertia matrices. Also suppose that the generalized velocities change as much as $\Delta\dot{\boldsymbol{\theta}}_i$ due to the impact forces $\boldsymbol{F}_i$ applied to the two chains at the connection point, and a new p-degrees-of-freedom joint $\boldsymbol{\theta}_n \in \boldsymbol{R}^p$ is created. According to the discussion in the previous subsection, a new virtual link is created at the connection point. Let $\boldsymbol{J}_n$ be the Jacobian matrix of the virtual link with respect to $\boldsymbol{\theta}_n$.

The applied force and the change of momentum of each chain are related by

$$\boldsymbol{A}_i \Delta\dot{\boldsymbol{\theta}}_i = -\boldsymbol{J}_i^T \boldsymbol{F}_i \;\; (i = 1, 2) \tag{3.1}$$

Since $\boldsymbol{F}_2$ is the reaction of $\boldsymbol{F}_1$, they satisfy

$$\boldsymbol{F}_2 = -\boldsymbol{F}_1. \tag{3.2}$$

On the other hand, the following equation is yielded by the condition that the velocity of the virtual link coincides with that of its real link.

$$\boldsymbol{J}_1(\dot{\boldsymbol{\theta}}_1 + \Delta\dot{\boldsymbol{\theta}}_1) = \boldsymbol{J}_2(\dot{\boldsymbol{\theta}}_2 + \Delta\dot{\boldsymbol{\theta}}_2) + \boldsymbol{J}_n\dot{\boldsymbol{\theta}}_n. \tag{3.3}$$

The impact force due to the collision has zero components along the free unconstrained directions of the new joint $\boldsymbol{\theta}_n$. This condition is expressed as

$$\boldsymbol{J}_n^T \boldsymbol{F}_1 = \boldsymbol{O}. \tag{3.4}$$

The change of generalized velocities $\Delta\dot{\boldsymbol{\theta}}_1$ and $\Delta\dot{\boldsymbol{\theta}}_2$ are computed from Eqs.(3.1)–(3.4) as

$$\Delta\dot{\boldsymbol{\theta}}_1 = -\boldsymbol{A}_1^{-1}\boldsymbol{J}_1^T\boldsymbol{B}^{-1}(\boldsymbol{I} - \boldsymbol{C})\boldsymbol{v} \tag{3.5}$$

$$\Delta\dot{\boldsymbol{\theta}}_2 = \boldsymbol{A}_2^{-1}\boldsymbol{J}_2^T\boldsymbol{B}^{-1}(\boldsymbol{I} - \boldsymbol{C})\boldsymbol{v} \tag{3.6}$$

where

$$\boldsymbol{B} = \boldsymbol{J}_1\boldsymbol{A}_1^{-1}\boldsymbol{J}_1^T + \boldsymbol{J}_2\boldsymbol{A}_2^{-1}\boldsymbol{J}_2^T \tag{3.7}$$

$$\boldsymbol{C} = \boldsymbol{J}_n(\boldsymbol{J}_n^T\boldsymbol{B}^{-1}\boldsymbol{J}_n)^{-1}\boldsymbol{J}_n^T\boldsymbol{B}^{-1} \tag{3.8}$$

$$\boldsymbol{v} = \boldsymbol{J}_1\dot{\boldsymbol{\theta}}_1 - \boldsymbol{J}_2\dot{\boldsymbol{\theta}}_2 \tag{3.9}$$

and $\boldsymbol{I}$ is a 6×6 identity matrix. If the connected two links are fixed to each other, namely, $p = 0$, $\Delta\dot{\boldsymbol{\theta}}_1$ and $\Delta\dot{\boldsymbol{\theta}}_2$ are computed by substituting $\boldsymbol{O}$ to $\boldsymbol{C}$ in Eqs.(3.5) and (3.6).

When the two links are in the same chain, the generalized coordinates and the mass matrices in the previous discussion coincide with each other, while the Jacobian matrices $\boldsymbol{J}_1$ and $\boldsymbol{J}_2$ are different. Therefore, the following equation is used in place of Eq.(3.1):

$$\boldsymbol{A}_1\Delta\dot{\boldsymbol{\theta}}_1 = -\boldsymbol{J}_1^T\boldsymbol{F}_1 - \boldsymbol{J}_2^T\boldsymbol{F}_2. \tag{3.10}$$

The unknown, $\Delta\dot{\boldsymbol{\theta}}_1$, is solved by

$$\Delta\dot{\boldsymbol{\theta}}_1 = -\boldsymbol{A}_1^{-1}\boldsymbol{J}_1^T\tilde{\boldsymbol{B}}^{-1}(\boldsymbol{I} - \tilde{\boldsymbol{C}})\tilde{\boldsymbol{v}} \tag{3.11}$$

where

$$\tilde{\boldsymbol{B}} = (\boldsymbol{J}_1 - \boldsymbol{J}_2)\boldsymbol{A}_1^{-1}(\boldsymbol{J}_1 - \boldsymbol{J}_2)^T \tag{3.12}$$

$$\tilde{\boldsymbol{C}} = \boldsymbol{J}_n(\boldsymbol{J}_n^T\tilde{\boldsymbol{B}}^{-1}\boldsymbol{J}_n)^{-1}\boldsymbol{J}_n^T\tilde{\boldsymbol{B}}^{-1} \tag{3.13}$$

$$\tilde{\boldsymbol{v}} = (\boldsymbol{J}_1 - \boldsymbol{J}_2)\dot{\boldsymbol{\theta}}_1. \tag{3.14}$$

3.5 Simulation Examples

We show two simulations of a simple human figure with structural changes. The human figure in the simulations has 16 degrees of freedom (4 for each arm and leg) as illustrated in Fig. 3.11. We applied zero torques except for the case when we need to restrict the joint angles within their limits. Computing the joint accelerations took approximately 25 ms on a Pentium Pro 200MHz processor in both examples.

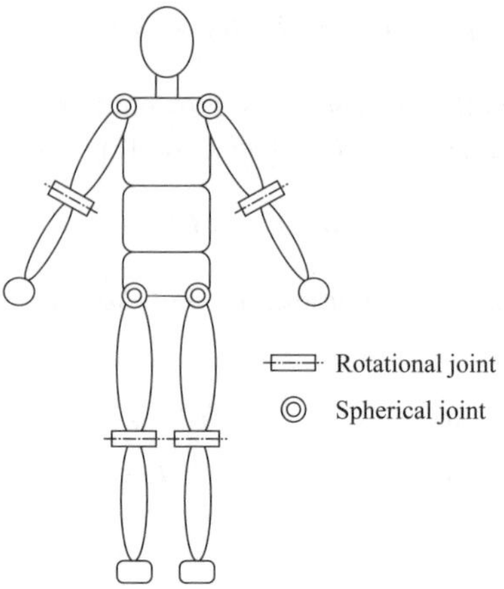

Fig. 3.11. 16-degrees-of-freedom human figure used in the simulations.

Fig. 3.12. High bar example of simulation of a human figure.

High Bar

Fig. 3.12 shows a human figure playing high bar and releasing the right hand during the motion. Initially there is a rotational joint between each hand and the bar, one of which is cut at an arbitrary given time. The connections are maintained, as discussed before, by two virtual links, whose real links are left and right hands. One of them is deleted when the cut occurs. The figure will be completely free-flying if the other virtual link is also deleted. In this case, including free joints and joints at the hands, we initially had 24 degrees of freedom in total and used 23 degrees of freedom after releasing the right hand.

Swing

What happens if a swing breaks down while you are playing on it? The answer is shown in Fig. 3.13. Each hand and the rod of the swing is connected by three-degrees-of-freedom spherical joint. There is a translational joint between each thigh and the plate of the swing, which is programmed to be cut when the thigh goes out of the plate. In this case, we initially had 30 degrees of freedom in total and used 28 degrees of freedom in the final figure of Fig. 3.13, including the swing.

Fig. 3.13. Swing example of simulation of a human figure.

3.6 Summary

This chapter presented a method for describing the link connectivity of open and closed kinematic chains and modifying the description according to structural changes. The contributions of this chapter are summarized by the following three points:

(1) Link connectivity description using pointers and virtual links was proposed. Pointers are convenient for implementing recursive algorithms because they directly indicate the neighboring links towards the end and root links. Closed loops are easily identified by the virtual links. They also indicate the virtually cut joints for the dynamics computation method presented in Chapter 2.

(2) Link connectivity maintenance scheme for structural changes was proposed. Link connections and joint cuts are always processed by adding and removing a virtual link, respectively. Following this scheme, the kinematic chain after a link connection is always represented as a closed kinematic chain even if it appears to be an open kinematic chain. Although handling a closed kinematic chain would increase the computational cost for the dynamics, the overhead due to the structural changes is smaller than inverting the parent-child relationships of the links to handle it as an open kinematic chain.

(3) A method for computing the velocity boundary condition after link connection was presented. The method computes the discontinuous change of the joint velocities due to the collision using the conservation of momentum and constraint conditions after the collision.

4

Parallel O(logN) Formulation of Forward Dynamics

4.1 Introduction

This chapter presents a forward dynamics formulation improved in terms of the asymptotic complexity compared to the method described in Chapter 2. Using the method in Chapter 2, the computational cost would increase rapidly as the DOF increases because its complexity is $O(N^3)$ where N denotes the degrees of freedom. The complexity of the new algorithm is as low as $O(N)$ complexity for all open kinematic chains and most of practical closed kinematic chains. The complexity is further reduced down to $O(\log N)$ by using $O(N)$ processors in parallel.

The formulation consists of two parts which can be interpreted as assembling and disassembling the target chain by adding and removing joints one by one, respectively. The computation starts from the initial state where all joints are removed and the links are not constrained at all. In the assembly phase, the joints are added one by one to finally complete the target chain. Each time a new joint is added, we compute the constraint force at the new joint and accelerations at the joints to be added in the following assembly procedures. Note that the quantities computed here is not valid for the complete chain since they may change due to the effect of joints added afterwards. This is why we need the disassembly phase, where the constraint forces and joint accelerations for the complete chain are computed as the joints are removed in the reverse order of the assembly phase.

The $O(N)$ complexity is realized by bounding the computational cost of each assembly and disassembly procedure. The computational cost of each step is determined by the number of joints through which an intermediate chain is connected to other chains. The number can be limited to three for all open kinematic chains and most practical closed ones by the aid of link splitting.

At each step of the two phases, if any pair of intermediate chains were independent of each other, assembling and disassembling those chains can be processed in

---

This chapter was adapted from, by permission, K. Yamane and Y. Nakamura, "Efficient Parallel Dynamics Computation of Human Figures," Proceedings of International Conference on Robotics and Automation, pp.530–537, Washington DC, May 2002.

parallel. It is also obvious that the parallelism of the formulation depend only on the order of adding and removing joints in the two phases. We do not have to prepare an algorithm customized for parallel computation. Generally speaking, increasing the parallelism of an algorithm leads to larger total computational cost and optimization of the parallelism for a specific target chain and the number of available processors is critical for the performance. One of the features of the new formulation is that we can tune the parallelism simply by re-ordering the assembly and disassembly phases.

This chapter is organized as follows. After a brief summary of existing efficient algorithms for dynamics simulation, Section 4.3 shows the idea of the new formulation. Section 4.4 describes the notations and some preliminaries used in the detailed algorithm in Section 4.5. Section 4.6 deals with the issues related to parallel computation such as scheduling and communication. Finally, we show several numerical examples in Section 4.9 using a 16-node PC cluster.

4.2 Related Work

Forward dynamics algorithms are often characterized by their asymptotic complexity. Letting N denote the degrees of freedom of the system, forward dynamics is an N-input (joint torques), N-output (joint accelerations) system. The theoretically lowest asymptotic complexity of forward dynamics computation is thus $O(N)$. Many algorithms have realized this complexity, including Articulated-Body Method [22], recursive dynamics formulations [89, 7], and efficient decomposition of coefficient matrix of the equation of motion using its sparsity [12].

Parallel computation is one way to accomplish even lower complexity. Fijany et al. [26] proposed the first forward dynamics algorithm with $O(\log N)$ asymptotic complexity for serial chain without branches and then extended it to allow short branches from the main chain [25]. Later, Featherstone [23, 24] proposed Divide-and-Conquer Method with the same complexity for a large set of kinematic chains, including closed ones. Hybrid Direct and Iterative Algorithm [3] combines exact and iterative solutions of equation of motion to parallelize the forward dynamics computation.

More discussion on the relationship between the assembly-disassembly algorithm and existing algorithms can be found in Section 4.8.

4.3 Overview

The idea of the algorithm is illustrated in Fig. 4.1. The free-flying links are assembled by adding a joint one by one to form the target kinematic chain, and then disassembled by removing the joints in the reverse order to return to the initial state. We refer to the intermediate kinematic chains found in the course of assembling as *subchains*. The target kinematic chain may be assembled in an arbitrary order. Therefore, if the subchains do not have connections with each others, we can process the assembly and disassembly computations in parallel. In Fig. 4.1, for example, the two joints

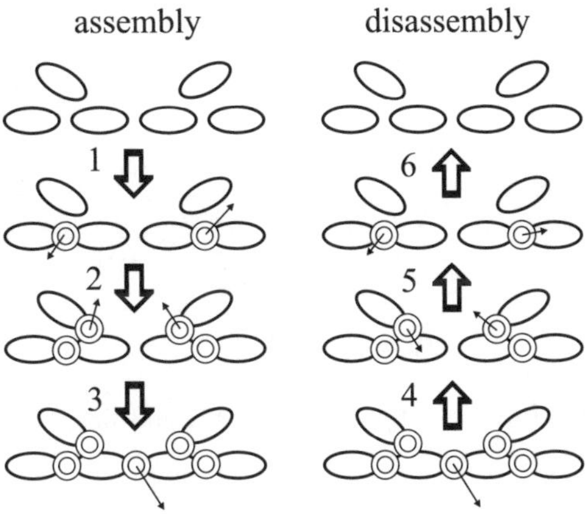

Fig. 4.1. The idea of assembling and disassembling a kinematic chain.

added or removed in steps 1, 2, 5 and 6 can be processed in parallel. Thus, although the kinematic chain contains 5 joints, the total computation time would be almost equivalent to handling 3 joints serially on one process.

The assembly phase computes two quantities: the constraint force at the new joint, and the acceleration of the points where new joints are to be added in the future. Note that the joints to be added in the future assembly steps are not considered at this stage. This is why we need the disassembly phase, where the constraint forces in the completed kinematic chains are computed. Once we have all constraint forces, the link accelerations are easily computed by applying Newton and Euler equations of motion to each link. The accelerations of two neighboring links are then used to compute the acceleration of the joint between them.

4.4 Preliminaries and Notations

4.4.1 Description of Link Connectivity

We employ the same link connectivity description described in Chapter 3 for implementation, i.e. the link connectivity is described by pointers and the closed loops by virtual links. For the presentation of the algorithm, however, we assign an index to each link and joint, including joints connecting virtual links and corresponding real links. The indices can be assigned arbitrarily as long as there is no duplicate within the indices of links or joints, except for the virtual links which are assigned the same indices as their real links. In this chapter, we treat a pair of real and virtual links as a single link.

The following variables and sets are defined for the indices (Fig. 4.2):

- p_i: represents the index of the link connected to the parent side of joint i

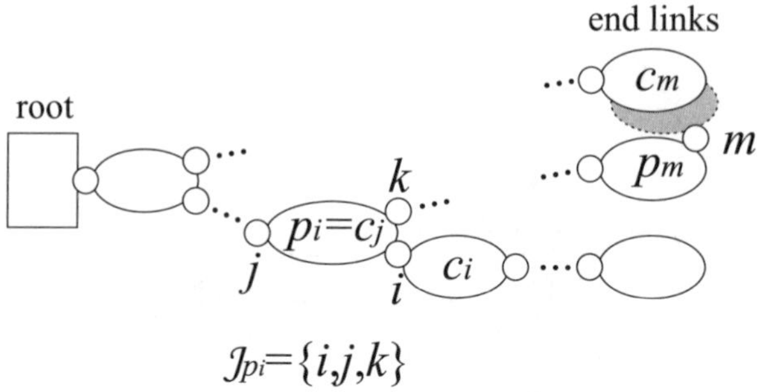

Fig. 4.2. Examples of p_i, c_i and $\mathcal{J}_k$. The gray link is a virtual.

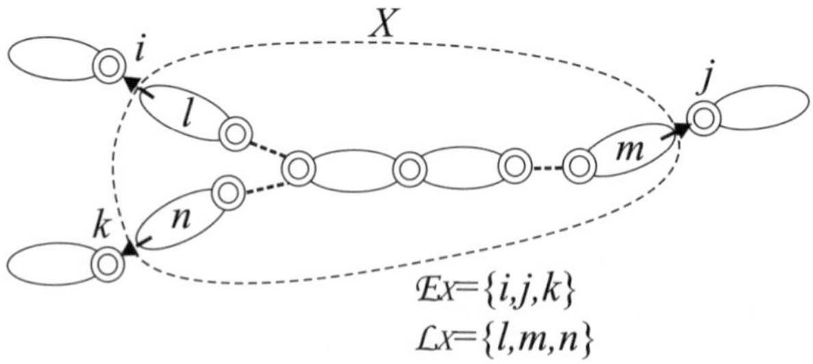

Fig. 4.3. Variables for subchain X.

- c_i: represents the index of the link connected to the child side (or the virtual link if the joint connects a virtual link) of joint i
- $\mathcal{J}_k$: the set that includes the indices of all joints connected to link k

We also assign a unique number to each subchain. The following variables are defined for subchain X (Fig. 4.3):

- N_{LX}: number of links
- N_{JX}: number of joints
- N_{CX}: total number of joint constraints
- N_{FX}: total number of DOF
- $\mathcal{E}_X$: set of indices of joints that connect subchain X to other subchains
- $\mathcal{L}_X$: set of links in subchain X that are connected to joints in $\mathcal{E}_X$

4.4.2 Frames

We define a frame for each link and two frames for each joint, as well as the fixed global frame (Fig. 4.4). The symbols for the frames are as follows:

- Σ_0: global frame
- Σ_k: frame attached to link k

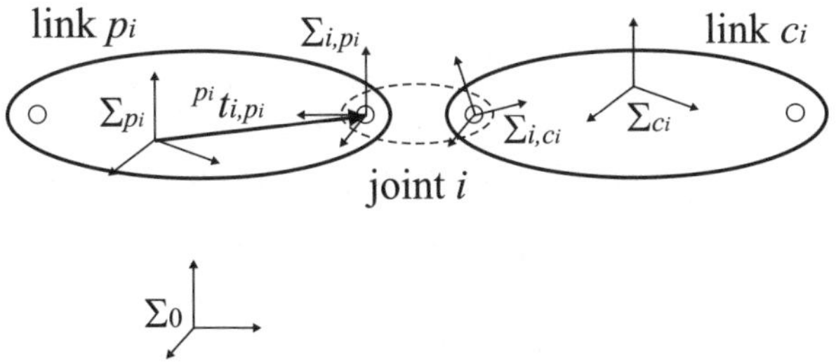

Fig. 4.4. Frames defined for links and joints.

- $\Sigma_{i,k}$: frame attached to joint i, fixed to link k ($k = \{p_i, c_i\}$) at relative position ${}^k\boldsymbol{t}_{i,k}$ and orientation ${}^k\boldsymbol{A}_{i,k}$

If a vector has a superscript on the left-hand side, it represents the frame in which the vector is described. In the following equations, $\boldsymbol{p}_*$, $\boldsymbol{R}_*$, $\boldsymbol{v}_*$, and $\boldsymbol{\omega}_*$ denote the position, orientation, linear velocity, and angular velocity of frame Σ_*, respectively. For example, ${}^0\boldsymbol{v}_{i,k}$ denotes the linear velocity of joint i frame attached to link k, described in the global frame.

4.4.3 Kinematic Equations

If joint i is attached to link k (that is, $k = p_i$ or $k = c_i$), the linear and angular velocities of frame $\Sigma_{i,k}$ are computed from those of frame Σ_k as

$$ {}^0\boldsymbol{v}_{i,k} = {}^0\boldsymbol{v}_k + {}^0\boldsymbol{\omega}_k \times {}^0\boldsymbol{t}_{i,k} \tag{4.1} $$

$$ {}^0\boldsymbol{\omega}_{i,k} = {}^0\boldsymbol{\omega}_k. \tag{4.2} $$

These equations are simplified as

$$ {}^0\dot{\boldsymbol{r}}_{i,k} = {}^0\boldsymbol{J}_{i,k}{}^0\dot{\boldsymbol{r}}_k \tag{4.3} $$

using combined vectors and matrices defined as

$$ {}^0\dot{\boldsymbol{r}}_{i,k} \triangleq \begin{pmatrix} {}^0\boldsymbol{v}_{i,k} \\ {}^0\boldsymbol{\omega}_{i,k} \end{pmatrix} \tag{4.4} $$

$$ {}^0\dot{\boldsymbol{r}}_k \triangleq \begin{pmatrix} {}^0\boldsymbol{v}_k \\ {}^0\boldsymbol{\omega}_k \end{pmatrix} \tag{4.5} $$

$$ {}^0\boldsymbol{J}_{i,k} \triangleq \begin{pmatrix} \mathbf{1}_3 & [{}^0\boldsymbol{t}_{i,k}\times]^T \\ \boldsymbol{O} & \mathbf{1}_3 \end{pmatrix} \tag{4.6} $$

where $[{}^0\boldsymbol{v}_{i,k}\times]$ is the cross product matrix of ${}^0\boldsymbol{v}_{i,k}$ and $\mathbf{1}_3$ is the 3×3 identity matrix. Converting ${}^0\dot{\boldsymbol{r}}_{i,k}$ to frame $\Sigma_{i,k}$ and ${}^0\dot{\boldsymbol{r}}_k$ to frame Σ_k, we obtain

$$ {}^{i,k}\dot{\boldsymbol{r}}_{i,k} = \boldsymbol{J}_{i,k}{}^k\dot{\boldsymbol{r}}_k \tag{4.7} $$

where

$$\boldsymbol{J}_{i,k} = \begin{pmatrix} {}^k\boldsymbol{A}_{i,k}^T & {}^k\boldsymbol{A}_{i,k}^T[{}^k\boldsymbol{t}_{i,k}\times]^T \\ \boldsymbol{O} & {}^k\boldsymbol{A}_{i,k}^T \end{pmatrix}. \tag{4.8}$$

Similarly, the relationship of linear and angular accelerations is written as

$${}^{i,k}\ddot{\boldsymbol{r}}_{i,k} = \boldsymbol{J}_{i,k}{}^k\ddot{\boldsymbol{r}}_k + \dot{\boldsymbol{\jmath}}_{i,k} \tag{4.9}$$

where

$$\dot{\boldsymbol{\jmath}}_{i,k} = \begin{pmatrix} {}^k\boldsymbol{A}_{i,k}^T({}^k\boldsymbol{\omega}_k \times ({}^k\boldsymbol{\omega}_k \times {}^k\boldsymbol{t}_{i,k})) \\ \boldsymbol{O} \end{pmatrix}. \tag{4.10}$$

4.4.4 Equation of Motion of a Link

The force and moment acting on the center of mass of link k is described as

$${}^0\boldsymbol{f}_k^* = m_k{}^0\dot{\boldsymbol{v}}_k^* \tag{4.11}$$

$${}^0\boldsymbol{n}_k^* = {}^0\boldsymbol{R}_k\boldsymbol{I}_k{}^0\boldsymbol{R}_k^{T0}\dot{\boldsymbol{\omega}}_k + {}^0\boldsymbol{\omega}_k \times ({}^0\boldsymbol{R}_k\boldsymbol{I}_k{}^0\boldsymbol{R}_k^{T0}\boldsymbol{\omega}_k) \tag{4.12}$$

where $\dot{\boldsymbol{v}}_k^*$ is the linear acceleration of the center of mass, m_k and $\boldsymbol{I}_k$ are the mass and inertia of link k respectively. ${}^0\boldsymbol{f}_k^*$ and ${}^0\boldsymbol{n}_k^*$ are converted to the force and moment acting on the origin of Σ_k by

$${}^0\boldsymbol{f}_k = {}^0\boldsymbol{f}_k^* \tag{4.13}$$

$${}^0\boldsymbol{n}_k = {}^0\boldsymbol{n}_k^* + {}^0\boldsymbol{s}_k \times {}^0\boldsymbol{f}_k^* \tag{4.14}$$

where ${}^0\boldsymbol{s}_k$ is the vector from the origin of Σ_k to the center of mass of link k. ${}^0\dot{\boldsymbol{v}}_k^*$ is computed from ${}^0\dot{\boldsymbol{v}}_k$ and ${}^0\dot{\boldsymbol{\omega}}_k$ as

$${}^0\dot{\boldsymbol{v}}_k^* = {}^0\dot{\boldsymbol{v}}_k + {}^0\dot{\boldsymbol{\omega}}_k \times {}^0\boldsymbol{s}_k + {}^0\boldsymbol{\omega}_k \times ({}^0\boldsymbol{\omega}_k \times {}^0\boldsymbol{s}_k) + {}^0\boldsymbol{g} \tag{4.15}$$

where ${}^0\boldsymbol{g}$ is the gravity acceleration, which is incorporated here to account for the gravitational force. Using Eqs.(4.13)–(4.15), we obtain the equation of motion of link k in the inertial frame:

$${}^0\boldsymbol{f}_k = m_k({}^0\dot{\boldsymbol{v}}_k + {}^0\dot{\boldsymbol{\omega}}_k \times {}^0\boldsymbol{s}_k + {}^0\boldsymbol{\omega}_k \times ({}^0\boldsymbol{\omega}_k \times {}^0\boldsymbol{s}_k) + {}^0\boldsymbol{g}) \tag{4.16}$$

$${}^0\boldsymbol{n}_k = {}^0\boldsymbol{R}_k\boldsymbol{I}_k{}^0\boldsymbol{R}_k^{T0}\dot{\boldsymbol{\omega}}_k + {}^0\boldsymbol{\omega}_k \times ({}^0\boldsymbol{R}_k\boldsymbol{I}_k{}^0\boldsymbol{R}_k^{T0}\boldsymbol{\omega}_k) + {}^0\boldsymbol{s}_k \times {}^0\boldsymbol{f}_k. \tag{4.17}$$

Converting these equations to frame Σ_k, we obtain

$${}^k\boldsymbol{\tau}_k = \boldsymbol{M}_k{}^k\ddot{\boldsymbol{r}}_k + \boldsymbol{c}_k \tag{4.18}$$

where

$${}^k\boldsymbol{\tau}_k = \begin{pmatrix} {}^k\boldsymbol{f}_k \\ {}^k\boldsymbol{n}_k \end{pmatrix} \tag{4.19}$$

$$\boldsymbol{M}_k = \begin{pmatrix} m_k\boldsymbol{1}_3 & m_k[{}^k\boldsymbol{s}_k\times]^T \\ m_k[{}^k\boldsymbol{s}_k\times] & \boldsymbol{I}_k - m_k[{}^k\boldsymbol{s}_k\times][{}^k\boldsymbol{s}_k\times] \end{pmatrix} \tag{4.20}$$

$$\boldsymbol{c}_k = \begin{pmatrix} m_k{}^k\boldsymbol{a}_k \\ {}^k\boldsymbol{\omega}_k \times (\boldsymbol{I}_k{}^k\boldsymbol{\omega}_k) + m_k{}^k\boldsymbol{s}_k \times {}^k\boldsymbol{a}_k \end{pmatrix} \tag{4.21}$$

$${}^k\boldsymbol{a}_k = {}^k\boldsymbol{\omega}_k \times ({}^k\boldsymbol{\omega}_k \times {}^k\boldsymbol{s}_k) + {}^0\boldsymbol{R}_k^{T0}\boldsymbol{g}. \tag{4.22}$$

4.4.5 Joint Constraints and Variables

A joint imposes some constraints on the relative motion of a pair of links. Suppose the constraints of joint i are described as

$$\boldsymbol{K}_{Ci}({}^{i,p_i}\dot{\boldsymbol{r}}_{i,c_i} - {}^{i,p_i}\dot{\boldsymbol{r}}_{i,p_i}) = \boldsymbol{O} \tag{4.23}$$

$\boldsymbol{K}_{Ci}$ is a $(6 - n_{Fi}) \times 6$ matrix, where n_{Fi} is the degrees of freedom of joint i, that projects the relative velocity between links p_i and c_i onto the constraint space. For a rotational joint that rotates around the z axis of Σ_{i,p_i}, for example, $\boldsymbol{K}_{Ci}$ would be

$$\boldsymbol{K}_{Ci} = \begin{pmatrix} 1\,0\,0\,0\,0\,0 \\ 0\,1\,0\,0\,0\,0 \\ 0\,0\,1\,0\,0\,0 \\ 0\,0\,0\,1\,0\,0 \\ 0\,0\,0\,0\,1\,0 \end{pmatrix}. \tag{4.24}$$

We can also define the joint velocities $\boldsymbol{q}_i$ as

$$\boldsymbol{q}_i = \boldsymbol{K}_{Ji}({}^{i,p_i}\dot{\boldsymbol{r}}_{i,c_i} - {}^{i,p_i}\dot{\boldsymbol{r}}_{i,p_i}) \tag{4.25}$$

where $\boldsymbol{K}_{Ji}$ is an $n_{Fi} \times 6$ matrix that projects the relative velocity onto the unconstrained space. For the same example as above, $\boldsymbol{K}_{Ji}$ would be

$$\boldsymbol{K}_{Ji} = \begin{pmatrix} 0\,0\,0\,0\,0\,1 \end{pmatrix}. \tag{4.26}$$

$\boldsymbol{K}_{Ci}$ and $\boldsymbol{K}_{Ji}$ are constant for most of the practical joint types such as rotational, prismatic, spherical, or universal joints. We assume they are constant for clarity of the explanation in this chapter, although it only requires straightforward changes to include time-dependent $\boldsymbol{K}_{Ci}$ and $\boldsymbol{K}_{Ji}$. Using this assumption, the acceleration constraints and joint accelerations are described as

$$\boldsymbol{K}_{Ci}({}^{i,p_i}\ddot{\boldsymbol{r}}_{i,c_i} - {}^{i,p_i}\ddot{\boldsymbol{r}}_{i,p_i}) = \boldsymbol{O} \tag{4.27}$$

$$\ddot{\boldsymbol{q}}_i = \boldsymbol{K}_{Ji}({}^{i,p_i}\ddot{\boldsymbol{r}}_{i,c_i} - {}^{i,p_i}\ddot{\boldsymbol{r}}_{i,p_i}). \tag{4.28}$$

We introduce new matrices for simple representation. Eqs.(4.23) and (4.25) are rewritten using Eq.(4.7) as

$$\boldsymbol{H}_{Ci,c_i}{}^{c_i}\dot{\boldsymbol{r}}_{c_i} + \boldsymbol{H}_{Ci,p_i}{}^{p_i}\dot{\boldsymbol{r}}_{p_i} = \boldsymbol{O} \tag{4.29}$$

$$\dot{\boldsymbol{q}}_i = \boldsymbol{H}_{Ji,c_i}{}^{c_i}\dot{\boldsymbol{r}}_{c_i} + \boldsymbol{H}_{Ji,p_i}{}^{p_i}\dot{\boldsymbol{r}}_{p_i} \tag{4.30}$$

where

$$\boldsymbol{H}_{Ci,c_i} = \boldsymbol{K}_{Ci}{}^{i,p_i}\hat{\boldsymbol{R}}_{i,c_i}\boldsymbol{J}_{i,c_i} \tag{4.31}$$

$$\boldsymbol{H}_{Ci,p_i} = -\boldsymbol{K}_{Ci}\boldsymbol{J}_{i,p_i} \tag{4.32}$$

$$\boldsymbol{H}_{Ji,c_i} = \boldsymbol{K}_{Ji}{}^{i,p_i}\hat{\boldsymbol{R}}_{i,c_i}\boldsymbol{J}_{i,c_i} \tag{4.33}$$

$$\boldsymbol{H}_{Ji,p_i} = -\boldsymbol{K}_{Ji}\boldsymbol{J}_{i,p_i} \tag{4.34}$$

$${}^{i,p_i}\hat{\boldsymbol{R}}_{i,c_i} = \begin{pmatrix} {}^{i,p_i}\boldsymbol{R}_{i,c_i} & \boldsymbol{O} \\ \boldsymbol{O} & {}^{i,p_i}\boldsymbol{R}_{i,c_i} \end{pmatrix}. \tag{4.35}$$

Similarly, Eqs.(4.27) and (4.28) are rewritten using Eq.(4.9) as

$$\boldsymbol{H}_{Ci,c_i}{}^{c_i}\ddot{\boldsymbol{r}}_{c_i} + \boldsymbol{H}_{Ci,p_i}{}^{p_i}\ddot{\boldsymbol{r}}_{p_i} + \boldsymbol{h}_{Ci,c_i} + \boldsymbol{h}_{Ci,p_i} = \boldsymbol{O} \tag{4.36}$$

$$\ddot{\boldsymbol{q}}_i = \boldsymbol{H}_{Ji,c_i}{}^{c_i}\ddot{\boldsymbol{r}}_{c_i} + \boldsymbol{H}_{Ji,p_i}{}^{p_i}\ddot{\boldsymbol{r}}_{p_i} + \boldsymbol{h}_{Ji,c_i} + \boldsymbol{h}_{Ji,p_i} \tag{4.37}$$

where

$$\boldsymbol{h}_{Ci,c_i} = \boldsymbol{K}_{Ci}{}^{i,p_i}\hat{\boldsymbol{R}}_{i,c_i}\dot{\boldsymbol{J}}_{i,c_i} \tag{4.38}$$

$$\boldsymbol{h}_{Ci,p_i} = -\boldsymbol{K}_{Ci}\dot{\boldsymbol{J}}_{i,p_i} \tag{4.39}$$

$$\boldsymbol{h}_{Ji,c_i} = \boldsymbol{K}_{Ji}{}^{i,p_i}\hat{\boldsymbol{R}}_{i,c_i}\dot{\boldsymbol{J}}_{i,c_i} \tag{4.40}$$

$$\boldsymbol{h}_{Ji,p_i} = -\boldsymbol{K}_{Ji}\dot{\boldsymbol{J}}_{i,p_i}. \tag{4.41}$$

Suppose link c_i receives the constraint force/torque $\boldsymbol{f}_{Ci} \in \boldsymbol{R}^{6-n_{Fi}}$ and the active joint force/torque $\boldsymbol{f}_{Ji} \in \boldsymbol{R}^{n_{Fi}}$ from joint i. These forces and torques are converted to the joint generalized force ${}^{i,p_i}\boldsymbol{f}_i$ by

$${}^{i,p_i}\boldsymbol{f}_i = \boldsymbol{K}_{Ci}^T\boldsymbol{f}_{Ci} + \boldsymbol{K}_{Ji}^T\boldsymbol{f}_{Ji}. \tag{4.42}$$

${}^{i,p_i}\boldsymbol{f}_i$ is then converted to frame Σ_{c_i} using $\boldsymbol{J}_{i,c_i}$ by

$${}^{c_i}\boldsymbol{\tau}_{c_i,i} = \boldsymbol{J}_{i,c_i}^T\,{}^{i,p_i}\hat{\boldsymbol{R}}_{i,c_i}^T\,{}^{i,p_i}\boldsymbol{f}_i \tag{4.43}$$

$$= \boldsymbol{H}_{Ci,c_i}^T\boldsymbol{f}_{Ci} + \boldsymbol{H}_{Ji,c_i}^T\boldsymbol{f}_{Ji} \tag{4.44}$$

where $\boldsymbol{\tau}_{k,i}$ denotes the contribution of joint i to the total generalized force acting on link k. Taking into account that link p_i receives $-{}^{i,p_i}\boldsymbol{f}_i$ from joint i, $\boldsymbol{\tau}_{p_i,i}$ is computed by

$${}^{p_i}\boldsymbol{\tau}_{p_i,i} = \boldsymbol{H}_{Ci,p_i}^T\boldsymbol{f}_{Ci} + \boldsymbol{H}_{Ji,p_i}^T\boldsymbol{f}_{Ji}. \tag{4.45}$$

Adding all the generalized forces from joints attached to link k, ${}^k\boldsymbol{\tau}_k$ is computed by

$${}^k\boldsymbol{\tau}_k = \sum_{i\in\mathcal{J}_k}{}^k\boldsymbol{\tau}_{k,i} \tag{4.46}$$

$${}^k\boldsymbol{\tau}_{k,i} = \boldsymbol{H}_{Ci,k}^T\boldsymbol{f}_{Ci} + \boldsymbol{H}_{Ji,k}^T\boldsymbol{f}_{Ji} \tag{4.47}$$

4.4.6 Summary of Basic Equations

Hereinafter, the vectors will be described in link or joint local frame, except when otherwise specified.

The relevant equations are summarized as follows:

- Equation of motion of link k

$$\boldsymbol{\tau}_k = \boldsymbol{M}_k\ddot{\boldsymbol{r}}_k + \boldsymbol{c}_k \tag{4.48}$$

- Joint constraints

$$\boldsymbol{H}_{Ci,c_i}\ddot{\boldsymbol{r}}_{c_i} + \boldsymbol{H}_{Ci,p_i}\ddot{\boldsymbol{r}}_{p_i} + \boldsymbol{h}_{Ci,c_i} + \boldsymbol{h}_{Ci,p_i} = \boldsymbol{O} \tag{4.49}$$

- Joint accelerations

$$\ddot{\boldsymbol{q}}_i = \boldsymbol{H}_{Ji,c_i}\ddot{\boldsymbol{r}}_{c_i} + \boldsymbol{H}_{Ji,p_i}\ddot{\boldsymbol{r}}_{p_i} + \boldsymbol{h}_{Ji,c_i} + \boldsymbol{h}_{Ji,p_i} \tag{4.50}$$

- Joint and constraint torques/forces

$$\boldsymbol{\tau}_k = \sum_{i \in \mathcal{J}_k} \boldsymbol{\tau}_{k,i} \tag{4.51}$$

$$\boldsymbol{\tau}_{k,i} = \boldsymbol{H}_{Ci,k}^T \boldsymbol{f}_{Ci} + \boldsymbol{H}_{Ji,k}^T \boldsymbol{f}_{Ji} \tag{4.52}$$

4.5 Details

4.5.1 Equation of Motion of a Subchain

Gathering Eq.(4.48) of all links in subchain X, we obtain

$$\boldsymbol{\tau}_X = \boldsymbol{M}_X \ddot{\boldsymbol{r}}_X + \boldsymbol{c}_X \tag{4.53}$$

where

$$\boldsymbol{\tau}_X = \begin{pmatrix} \boldsymbol{\tau}_{l_{X1}} \\ \boldsymbol{\tau}_{l_{X2}} \\ \vdots \\ \boldsymbol{\tau}_{l_{XN_{LX}}} \end{pmatrix}$$

$$\boldsymbol{M}_X = \text{diag}(\boldsymbol{M}_{l_{Xk}})$$

$$\ddot{\boldsymbol{r}}_X = \begin{pmatrix} \ddot{\boldsymbol{r}}_{l_{X1}} \\ \ddot{\boldsymbol{r}}_{l_{X2}} \\ \vdots \\ \ddot{\boldsymbol{r}}_{l_{XN_{LX}}} \end{pmatrix}$$

$$\boldsymbol{c}_X = \begin{pmatrix} \boldsymbol{c}_{l_{X1}} \\ \boldsymbol{c}_{l_{X2}} \\ \vdots \\ \boldsymbol{c}_{l_{XN_{LX}}} \end{pmatrix}$$

and l_{Xk} $(k = 1 \ldots N_{LX})$ are the indices of the links in subchain X.

On the other hand, combining Eq.(4.49) of all joints in subchain X yields

$$\boldsymbol{H}_{CX}\ddot{\boldsymbol{r}}_X + \boldsymbol{h}_{CX} = \boldsymbol{O} \tag{4.54}$$

where $\boldsymbol{H}_{CX} \in \boldsymbol{R}^{N_{CX} \times 6N_{LX}}$ is a block matrix in the following form:

$$\boldsymbol{H}_{CX} = \begin{matrix} \\ i \\ \\ \end{matrix} \begin{pmatrix} & & \overset{p_{j_i}}{\vdots} & & \overset{c_{j_i}}{\vdots} & \\ \boldsymbol{O} \ldots \boldsymbol{O} & \boldsymbol{H}_{Cj_i,p_{j_i}} & & \boldsymbol{O} \ldots \boldsymbol{O} & \boldsymbol{H}_{Cj_i,c_{j_i}} & \boldsymbol{O} \ldots \boldsymbol{O} \\ & \vdots & & & \vdots & \end{pmatrix} \tag{4.55}$$

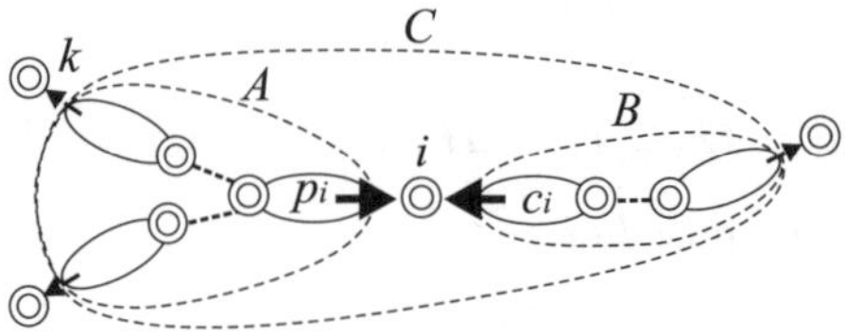

Fig. 4.5. Assembling two subchains.

namely, the i-th row has two nonzero blocks at columns p_{j_i} and c_{j_i} where j_i $(i = 1 \ldots N_{JX})$ is the indices of the joints in subchain X. $\boldsymbol{h}_{CX} \in \boldsymbol{R}^{N_{CX}}$ is a vector in the following form:

$$\boldsymbol{h}_{CX} = i \begin{pmatrix} \vdots \\ \boldsymbol{h}_{Cj_i,c_{j_i}} + \boldsymbol{h}_{Cj_i,p_{j_i}} \\ \vdots \end{pmatrix}. \tag{4.56}$$

The global form of Eq.(4.52) is described as

$$\boldsymbol{\tau}_X = \boldsymbol{H}_{CX}^T \boldsymbol{f}_{CX} + \boldsymbol{H}_{JX}^T \boldsymbol{f}_{JX} \tag{4.57}$$

where $\boldsymbol{H}_{JX}$ is formed in the same way as $\boldsymbol{H}_{CX}$ using $\boldsymbol{H}_{Ji,k}$.

Eqs.(4.48)(4.54) and (4.57) are solved in terms of the constraint forces $\boldsymbol{f}_{CX}$ as

$$\boldsymbol{f}_{CX} = \boldsymbol{S}_X^{-1}(-\boldsymbol{H}_{CX}\boldsymbol{M}_X^{-1}\boldsymbol{H}_{JX}^T\boldsymbol{\tau}_X + \boldsymbol{H}_{CX}\boldsymbol{M}_X^{-1}\boldsymbol{c}_X - \boldsymbol{h}_{CX}) \tag{4.58}$$

where

$$\boldsymbol{S}_X \triangleq \boldsymbol{H}_{CX}\boldsymbol{M}_X^{-1}\boldsymbol{H}_{CX}^T. \tag{4.59}$$

Evaluating Eq.(4.58) directly as in most of commercial software leads to an $O(N^2)$ algorithm, while making use of the sparsity of $\boldsymbol{S}_X$ yields more efficient $O(N)$ solution [12].

Since $\ddot{\boldsymbol{r}}_k$ and $\boldsymbol{f}_{Ci}$ change as the assembly phase proceeds, we denote the subchain in which the value is valid by the superscript on the left-hand side. A vector without the superscript is valid in the completed chain.

4.5.2 Assembly Phase

Suppose we are going to assemble subchains A and B by connecting links p_i and c_i through joint i to build subchain C as illustrated in Fig. 4.5.

Before adding joint i, the equations of motion and kinematic constraints of subchain X $(X = A, B)$ are

$$\boldsymbol{M}_X {}^X\ddot{\boldsymbol{r}}_X + \boldsymbol{c}_X = \boldsymbol{H}_{CX}^T {}^X\boldsymbol{f}_{CX} + \boldsymbol{H}_{JX}^T \boldsymbol{f}_{JX} \tag{4.60}$$

$$\boldsymbol{H}_{CX} {}^X\ddot{\boldsymbol{r}}_X + \boldsymbol{h}_{CX} = \boldsymbol{O}. \tag{4.61}$$

${}^A\ddot{\boldsymbol{r}}_{i,p_i}$ and ${}^B\ddot{\boldsymbol{r}}_{i,c_i}$ are computed by

$$ {}^A\ddot{\boldsymbol{r}}_{i,p_i} = \boldsymbol{J}_{i,A}{}^A\ddot{\boldsymbol{r}}_A + \dot{\boldsymbol{j}}_{i,p_i} \tag{4.62} $$

$$ {}^B\ddot{\boldsymbol{r}}_{i,c_i} = \boldsymbol{J}_{i,B}{}^B\ddot{\boldsymbol{r}}_B + \dot{\boldsymbol{j}}_{i,c_i} \tag{4.63} $$

where $\boldsymbol{J}_{i,A} \in \boldsymbol{R}^{6\times 6N_{LA}}$ and $\boldsymbol{J}_{i,B} \in \boldsymbol{R}^{6\times 6N_{LB}}$ are block matrices in the following forms:

$$ \boldsymbol{J}_{i,A} = \left(\boldsymbol{O}\ldots\boldsymbol{O}\ \boldsymbol{J}_{i,p_i}\ \boldsymbol{O}\ldots\boldsymbol{O}\right) \tag{4.64} $$

$$ \boldsymbol{J}_{i,B} = \left(\boldsymbol{O}\ldots\boldsymbol{O}\ \boldsymbol{J}_{i,c_i}\ \boldsymbol{O}\ldots\boldsymbol{O}\right) \tag{4.65} $$

namely each of $\boldsymbol{J}_{i,A}$ and $\boldsymbol{J}_{i,B}$ has one nonzero block.

From Eqs.(4.60)(4.61), the constraint forces of subchain A are computed by

$$ {}^A\boldsymbol{f}_{CA} = \boldsymbol{S}_A^{-1}(-\boldsymbol{T}_A\boldsymbol{f}_{JA} + \boldsymbol{H}_{CA}\boldsymbol{M}_A^{-1}\boldsymbol{c}_A - \boldsymbol{h}_{CA}) \tag{4.66} $$

where

$$ \boldsymbol{T}_A \triangleq \boldsymbol{H}_{CA}\boldsymbol{M}_A^{-1}\boldsymbol{H}_{JA}^T. \tag{4.67} $$

Similarly for subchain B, ${}^B\boldsymbol{f}_{CB}$ is computed by

$$ {}^B\boldsymbol{f}_{CB} = \boldsymbol{S}_B^{-1}(-\boldsymbol{T}_B\boldsymbol{f}_{JB} + \boldsymbol{H}_{CB}\boldsymbol{M}_B^{-1}\boldsymbol{c}_B - \boldsymbol{h}_{CB}) \tag{4.68} $$

where $\boldsymbol{T}_B$ is defined in the same way as in subchain A.

When we connect subchains A and B through joint i, we have the equations of motion

$$ \boldsymbol{M}_A{}^C\ddot{\boldsymbol{r}}_A + \boldsymbol{c}_A = \boldsymbol{H}_{CA}^T{}^C\boldsymbol{f}_{CA} + \boldsymbol{H}_{JA}^T\boldsymbol{f}_{JA} + \boldsymbol{H}_{Ci,A}^T{}^C\boldsymbol{f}_{Ci} + \boldsymbol{H}_{Ji,A}^T\boldsymbol{f}_{Ji} \tag{4.69} $$

$$ \boldsymbol{M}_B{}^C\ddot{\boldsymbol{r}}_B + \boldsymbol{c}_B = \boldsymbol{H}_{CB}^T{}^C\boldsymbol{f}_{CB} + \boldsymbol{H}_{JB}^T\boldsymbol{f}_{JB} + \boldsymbol{H}_{Ci,B}^T{}^C\boldsymbol{f}_{Ci} + \boldsymbol{H}_{Ji,B}^T\boldsymbol{f}_{Ji} \tag{4.70} $$

and the constraint conditions

$$ \boldsymbol{H}_{CA}{}^C\ddot{\boldsymbol{r}}_A + \boldsymbol{h}_{CA} = \boldsymbol{O} \tag{4.71} $$

$$ \boldsymbol{H}_{CB}{}^C\ddot{\boldsymbol{r}}_B + \boldsymbol{h}_{CB} = \boldsymbol{O} \tag{4.72} $$

$$ \boldsymbol{H}_{Ci,A}{}^C\ddot{\boldsymbol{r}}_A + \boldsymbol{h}_{Ci,p_i} + \boldsymbol{H}_{Ci,B}{}^C\ddot{\boldsymbol{r}}_B + \boldsymbol{h}_{Ci,c_i} = \boldsymbol{O} \tag{4.73} $$

where $\boldsymbol{H}_{Ci,A} \in \boldsymbol{R}^{(6-n_{F_i})\times 6N_{LA}}$ and $\boldsymbol{H}_{Ci,B} \in \boldsymbol{R}^{(6-n_{F_i})\times 6N_{LB}}$ have the following structure:

$$ \boldsymbol{H}_{Ci,A} = \left(\boldsymbol{O}\ldots\boldsymbol{O}\ \boldsymbol{H}_{Ci,p_i}\ \boldsymbol{O}\ldots\boldsymbol{O}\right) \tag{4.74} $$

$$ \boldsymbol{H}_{Ci,B} = \left(\boldsymbol{O}\ldots\boldsymbol{O}\ \boldsymbol{H}_{Ci,c_i}\ \boldsymbol{O}\ldots\boldsymbol{O}\right). \tag{4.75} $$

Combining Eqs.(4.69)(4.70) and Eqs.(4.71)–(4.73), we obtain

$$ \boldsymbol{M}_C{}^C\ddot{\boldsymbol{r}}_C + \boldsymbol{c}_C = \boldsymbol{H}_{CC}^T{}^C\boldsymbol{f}_{CC} + \boldsymbol{H}_{JC}^T\boldsymbol{f}_{JC} \tag{4.76} $$

$$ \boldsymbol{H}_{CC}{}^C\ddot{\boldsymbol{f}}_{CC} + \boldsymbol{h}_{CC} = \boldsymbol{O} \tag{4.77} $$

where

$$\boldsymbol{M}_C \triangleq \begin{pmatrix} \boldsymbol{M}_A & \boldsymbol{O} \\ \boldsymbol{O} & \boldsymbol{M}_B \end{pmatrix} \tag{4.78}$$

$${}^C\ddot{\boldsymbol{r}}_C \triangleq \begin{pmatrix} {}^C\ddot{\boldsymbol{r}}_A \\ {}^C\ddot{\boldsymbol{r}}_B \end{pmatrix} \tag{4.79}$$

$$\boldsymbol{c}_C \triangleq \begin{pmatrix} \boldsymbol{c}_A \\ \boldsymbol{c}_B \end{pmatrix} \tag{4.80}$$

$${}^C\boldsymbol{f}_{CC} \triangleq \begin{pmatrix} {}^C\boldsymbol{f}_{CA} \\ {}^C\boldsymbol{f}_{CB} \\ {}^C\boldsymbol{f}_{Ci} \end{pmatrix} \tag{4.81}$$

$$\boldsymbol{f}_{JC} \triangleq \begin{pmatrix} \boldsymbol{f}_{JA} \\ \boldsymbol{f}_{JB} \\ \boldsymbol{f}_{Ji} \end{pmatrix} \tag{4.82}$$

$$\boldsymbol{H}_{CC} \triangleq \begin{pmatrix} \boldsymbol{H}_{CA} & \boldsymbol{O} \\ \boldsymbol{O} & \boldsymbol{H}_{CB} \\ \boldsymbol{H}_{Ci,A} & \boldsymbol{H}_{Ci,B} \end{pmatrix} \tag{4.83}$$

$$\boldsymbol{h}_{CC} \triangleq \begin{pmatrix} \boldsymbol{h}_{CA} \\ \boldsymbol{h}_{CB} \\ \boldsymbol{h}_{i,p_i} + \boldsymbol{h}_{i,c_i} \end{pmatrix} \tag{4.84}$$

Solving Eqs. (4.69)–(4.73) in terms of ${}^C\boldsymbol{f}_i$ and simplifying it using Eqs. (4.62) (4.63)(4.66)(4.68) yields

$${}^C\boldsymbol{f}_{Ci} = -\boldsymbol{\Gamma}_{i,i}^{-1}\boldsymbol{K}_{Ci}({}^A\ddot{\boldsymbol{r}}_{i,p_i} + {}^B\ddot{\boldsymbol{r}}_{i,c_i} + \boldsymbol{P}_{i,i}\boldsymbol{K}_{Ji}^T\boldsymbol{f}_{Ji}) \tag{4.85}$$

where

$$\boldsymbol{\Gamma}_{i,i} = \boldsymbol{K}_{Ci}\boldsymbol{P}_{i,i}\boldsymbol{K}_{Ci}^T \tag{4.86}$$
$$\boldsymbol{P}_{i,i} = \boldsymbol{\Lambda}_{Ai,i} + \boldsymbol{\Lambda}_{Bi,i} \tag{4.87}$$
$$\boldsymbol{\Lambda}_{Xi,i} = \boldsymbol{J}_{i,X}\boldsymbol{\Phi}_X\boldsymbol{J}_{i,X}^T \tag{4.88}$$
$$\boldsymbol{\Phi}_X = \boldsymbol{M}_X^{-1} - \boldsymbol{M}_X^{-1}\boldsymbol{H}_{CX}^T\boldsymbol{S}_X^{-1}\boldsymbol{H}_{CX}\boldsymbol{M}_X^{-1} \tag{4.89}$$
$$\boldsymbol{S}_X = \boldsymbol{H}_{CX}\boldsymbol{M}_X^{-1}\boldsymbol{H}_{CX}^T. \tag{4.90}$$
$$(X = A, B) \tag{4.91}$$

Because ${}^X\ddot{\boldsymbol{r}}_{i,X}$ and $\boldsymbol{\Lambda}_{Xi,i}(X = A, B)$ of the subchains are required to compute ${}^C\boldsymbol{f}_{Ci}$, we have to compute ${}^C\ddot{\boldsymbol{r}}_{k,C}$ $(k \in \mathcal{E}_C)$ and $\boldsymbol{\Lambda}_{Cj,j}$ for the coming assembly computations, where j is the joint going to be added next. First, using the relationship

$$\ddot{\boldsymbol{r}}_{k,C} = \boldsymbol{J}_{k,C}\ddot{\boldsymbol{r}}_C + \dot{\boldsymbol{j}}_{k,C}, \tag{4.92}$$

${}^C\ddot{\boldsymbol{r}}_{k,C}$ is computed by

$${}^C\ddot{\boldsymbol{r}}_{k,C} = {}^X\ddot{\boldsymbol{r}}_{k,X} + \boldsymbol{\Lambda}_{Xk,i}(\boldsymbol{K}_{Ci}^T\,{}^C\boldsymbol{f}_{Ci} + \boldsymbol{K}_{Ji}^T\boldsymbol{f}_{Ji}) \tag{4.93}$$

where $\boldsymbol{\Lambda}_{Xk,i} = \boldsymbol{J}_{k,X}\boldsymbol{\Phi}_X\boldsymbol{J}_{i,X}^T$ and $X \in \{A, B\}$ is selected so that k is also included in $\mathcal{E}_X$.

Next, we compute $\boldsymbol{\Lambda}_{Cm,j}(m \in \mathcal{E}_C)$ which includes all of the $\boldsymbol{\Lambda}_C$ matrices required for the next assembly. Now we have

$$\boldsymbol{\Lambda}_{Cm,j} = \boldsymbol{J}_{m,C}\boldsymbol{\Phi}_C\boldsymbol{J}_{j,C}^T \tag{4.94}$$

$$\boldsymbol{\Phi}_C = \boldsymbol{M}_C^{-1} - \boldsymbol{M}_C^{-1}\boldsymbol{H}_{CC}^T\boldsymbol{S}_C^{-1}\boldsymbol{H}_{CC}\boldsymbol{M}_C^{-1} \tag{4.95}$$

$$\boldsymbol{S}_C = \boldsymbol{H}_{CC}\boldsymbol{M}_C^{-1}\boldsymbol{H}_{CC}^T. \tag{4.96}$$

We first yield a simplified expression of $\boldsymbol{\Phi}_C$. Using Eqs.(4.83)(4.78) and (4.96), $\boldsymbol{S}_C$ is written as

$$\boldsymbol{S}_C = \begin{pmatrix} \boldsymbol{S}_A & \boldsymbol{O} & \boldsymbol{S}_{Ai} \\ \boldsymbol{O} & \boldsymbol{S}_B & \boldsymbol{S}_{Bi} \\ \boldsymbol{S}_{Ai}^T & \boldsymbol{S}_{Bi}^T & \boldsymbol{S}_{ii} \end{pmatrix} \tag{4.97}$$

where

$$\boldsymbol{S}_{Ai} \stackrel{\triangle}{=} \boldsymbol{H}_{CA}\boldsymbol{M}_A^{-1}\boldsymbol{H}_{Ci,A}^T$$
$$\boldsymbol{S}_{Bi} \stackrel{\triangle}{=} -\boldsymbol{H}_{CB}\boldsymbol{M}_B^{-1}\boldsymbol{H}_{Ci,B}^T$$
$$\boldsymbol{S}_{ii} \stackrel{\triangle}{=} \boldsymbol{H}_{Ci,A}\boldsymbol{M}_A^{-1}\boldsymbol{H}_{Ci,A}^T + \boldsymbol{H}_{Ci,B}\boldsymbol{M}_B^{-1}\boldsymbol{H}_{Ci,B}^T.$$

which yields

$$\boldsymbol{S}_C^{-1} = \begin{pmatrix} \boldsymbol{S}_{CAA} & \boldsymbol{S}_{CAB} & \boldsymbol{S}_{CAi} \\ \boldsymbol{S}_{CAB}^T & \boldsymbol{S}_{CBB} & \boldsymbol{S}_{CBi} \\ \boldsymbol{S}_{CAi}^T & \boldsymbol{S}_{CBi}^T & \boldsymbol{S}_{Cii} \end{pmatrix} \tag{4.98}$$

where

$$\boldsymbol{S}_{CAA} = \boldsymbol{S}_A^{-1} + \boldsymbol{S}_A^{-1}\boldsymbol{S}_{Ai}\boldsymbol{\Gamma}_{i,i}^{-1}\boldsymbol{S}_{Ai}^T\boldsymbol{S}_A^{-1}$$
$$\boldsymbol{S}_{CAB} = \boldsymbol{S}_A^{-1}\boldsymbol{S}_{Ai}\boldsymbol{\Gamma}_{i,i}^{-1}\boldsymbol{S}_{Bi}^T\boldsymbol{S}_B^{-1}$$
$$\boldsymbol{S}_{CAi} = -\boldsymbol{S}_A^{-1}\boldsymbol{S}_{Ai}\boldsymbol{\Gamma}_{i,i}^{-1}$$
$$\boldsymbol{S}_{CBB} = \boldsymbol{S}_B^{-1} + \boldsymbol{S}_B^{-1}\boldsymbol{S}_{Bi}\boldsymbol{\Gamma}_{i,i}^{-1}\boldsymbol{S}_{Bi}^T\boldsymbol{S}_B^{-1}$$
$$\boldsymbol{S}_{CBi} = -\boldsymbol{S}_B^{-1}\boldsymbol{S}_{Bi}\boldsymbol{\Gamma}_{i,i}^{-1}$$
$$\boldsymbol{S}_{Cii} = \boldsymbol{\Gamma}_{i,i}^{-1}.$$

Substituting Eq.(4.98) into Eq.(4.95) and simplifying it using $\boldsymbol{\Phi}_A$ and $\boldsymbol{\Phi}_B$, we obtain

$$\boldsymbol{\Phi}_C = \begin{pmatrix} \boldsymbol{\Phi}_A - \boldsymbol{\Phi}_A\boldsymbol{H}_{Ci,A}^T\boldsymbol{\Gamma}_{i,i}^{-1}\boldsymbol{H}_{Ci,A}\boldsymbol{\Phi}_A & -\boldsymbol{\Phi}_A\boldsymbol{H}_{Ci,A}^T\boldsymbol{\Gamma}_{i,i}^{-1}\boldsymbol{H}_{Ci,B}\boldsymbol{\Phi}_B \\ -\boldsymbol{\Phi}_B\boldsymbol{H}_{Ci,B}^T\boldsymbol{\Gamma}_{i,i}^{-1}\boldsymbol{H}_{Ci,A}\boldsymbol{\Phi}_A & \boldsymbol{\Phi}_B - \boldsymbol{\Phi}_B\boldsymbol{H}_{Ci,B}^T\boldsymbol{\Gamma}_{i,i}^{-1}\boldsymbol{H}_{Ci,B}\boldsymbol{\Phi}_B \end{pmatrix}. \tag{4.99}$$

Computing all elements of $\boldsymbol{\Phi}_C$ requires $O(N^2)$ computations. Our final goal is, however, to evaluate $\boldsymbol{\Lambda}_{Cm,j}$ that requires only selected blocks of $\boldsymbol{\Phi}_C$ thanks to the structure of $\boldsymbol{J}_{m,C}$:

$$\boldsymbol{J}_{m,C} = \left(\boldsymbol{O} \ldots \boldsymbol{O}\ \boldsymbol{J}_{m,p_m}\ \boldsymbol{O} \ldots \boldsymbol{O} \right) \tag{4.100}$$

where $p_m \in \mathcal{L}_C$ is the link in subchain C that is connected to joint m. Therefore, we only need to compute the (p_m, p_j)-elements of $\boldsymbol{\Phi}_C$ because $\boldsymbol{\Lambda}_{Cm,j}$ is computed by

$$\boldsymbol{\Lambda}_{Cm,j} = \boldsymbol{J}_{m,p_m} \boldsymbol{\Phi}_{Cp_m,p_j} \boldsymbol{J}_{j,p_j}. \tag{4.101}$$

There are the following 4 possibilities in the placement of joints m and j:

(1) $m \in \mathcal{E}_A$ and $j \in \mathcal{E}_A$
(2) $m \in \mathcal{E}_A$ and $j \in \mathcal{E}_B$
(3) $m \in \mathcal{E}_B$ and $j \in \mathcal{E}_A$
(4) $m \in \mathcal{E}_B$ and $j \in \mathcal{E}_B$

In case (1), for example, we use the upper-left block of Eq.(4.99):

$$\begin{aligned} \boldsymbol{\Lambda}_{Cm,j} &= \boldsymbol{J}_{m,p_m} (\boldsymbol{\Phi}_A - \boldsymbol{\Phi}_A \boldsymbol{H}_{Ci,A}^T \boldsymbol{\Gamma}_{i,i}^{-1} \boldsymbol{H}_{Ci,A} \boldsymbol{\Phi}_A) \boldsymbol{J}_{j,p_j} \\ &= \boldsymbol{J}_{m,p_m} \boldsymbol{\Phi}_A \boldsymbol{J}_{j,p_j} - \boldsymbol{J}_{m,p_m} \boldsymbol{\Phi}_A \boldsymbol{J}_{i,A}^T \boldsymbol{K}_{Ci}^T \boldsymbol{\Gamma}_{i,i}^{-1} \boldsymbol{K}_{Ci} \boldsymbol{J}_{i,A} \boldsymbol{\Phi}_A \boldsymbol{J}_{j,p_j} \\ &= \boldsymbol{\Lambda}_{Am,j} - \boldsymbol{\Lambda}_{Am,i} \boldsymbol{K}_{Ci}^T \boldsymbol{\Gamma}_{i,i}^{-1} \boldsymbol{K}_{Ci} \boldsymbol{\Lambda}_{Ai,j} \end{aligned} \tag{4.102}$$

which only uses the quantities computed in the assembly process for constructing subchain A. Cases (2) to (4) are also handled in similar ways.

Computations for assembling subchain C through joint i consist of following four steps:

(1) compute $\boldsymbol{P}_{i,i}$ and $\boldsymbol{\Gamma}_{i,i}^{-1}$ (Eqs.(4.87) and (4.86)),
(2) compute ${}^C\boldsymbol{f}_{Ci}$ (Eq.(4.85)),
(3) compute $\boldsymbol{\Lambda}_{Cm,n}$ $(m, n \in \mathcal{E}_C)$ (Eq.(4.102)), and
(4) compute ${}^C\ddot{\boldsymbol{r}}_{m,C}$ $(m \in \mathcal{E}_C)$ (Eq.(4.93)).

4.5.3 Disassembly Phase

The constraint forces computed in the assembly phase are valid only in the corresponding subchains. After the completion of the assembly phase, the values might have changed due to the effects of joints added afterward. The disassembly phase computes the constraint forces in the completed chain by disassembling the subchains in the reverse order of the assembly phase. When a joint is removed, its final constraint force is computed, which in turn can be regarded as an external force for the two subchains which the joint had connected.

Suppose we are about to remove joint i. Joints in $\mathcal{E}_C$ were assembled *after* joint i; therefore, they are removed *before* joint i in the disassembly phase and we already know the final constraint forces of joint $k \in \mathcal{E}_C$, denoted by $\boldsymbol{f}_{Ck}$.

Regarding $\boldsymbol{f}_{Ck}$ $(k \in \mathcal{E}_C)$ as external forces, we have the new equations of motions for subchains A and B:

$$
\begin{aligned}
M_A\ddot{r}_A + c_A &= H_{Ci,A}^T f_{Ci} + H_{CA}^T f_{CA} + H_{Ji,A}^T f_{Ji} + H_{JA}^T f_{JA} \\
&+ \sum_{k\in\mathcal{E}_A} (H_{Ck,A}^T f_{Ck} + H_{Jk,A}^T f_{Jk}) \\
M_B\ddot{r}_B + c_B &= H_{Ci,B}^T f_{Ci} + H_{CB}^T f_{CB} + H_{Ji,B}^T f_{Ji} + H_{JB}^T f_{JB} \\
&+ \sum_{k\in\mathcal{E}_B} (H_{Ck,B}^T f_{Ck} + H_{Jk,B}^T f_{Jk})
\end{aligned}
$$

and the equations of constraints:

$$
\begin{aligned}
H_{CA}\ddot{r}_A + h_{CA} &= O \\
H_{CB}\ddot{r}_B + h_{CB} &= O \\
H_{Ci,A}\ddot{r}_A + h_{Ci,p_i} + H_{Ci,B}\ddot{r}_B + h_{Ci,c_i} &= O
\end{aligned}
$$

where the unknowns are $f_{Ci}, f_{CA}, f_{CB}, \ddot{r}_A$ and $\ddot{r}_B$. Solving these equations in terms of f_{Ci}, the final constraint force is computed by

$$
f_{Ci} = {}^C f_{Ci} - \Gamma_{i,i}^{-1} K_{Ci} \sum \Lambda_{Xk,i}^T (K_{Ck}^T f_{Ck} + K_{Jk}^T f_{Jk}). \tag{4.103}
$$

Once the constraint force is computed, the accelerations of the both sides of the joint are computed by

$$
\begin{aligned}
\ddot{r}_{i,X} &= {}^X\ddot{r}_{i,X} + \Lambda_{Xi,i}(K_{Ci}^T f_{Ci} + K_{Ji}^T f_{Ji}) \\
&+ \sum \Lambda_{Xk,i}^T (K_{Ck}^T f_{Ck} + K_{Jk}^T f_{Ck}) \\
&(X = A, B)
\end{aligned} \tag{4.104}
$$

where ${}^A\ddot{r}_{i,A} \triangleq {}^A\ddot{r}_{i,p_i}$ and ${}^B\ddot{r}_{i,B} \triangleq {}^B\ddot{r}_{i,c_i}$. Finally, the joint acceleration is computed by

$$
\ddot{q}_i = K_{Ji}({}^{i,p_i}\ddot{r}_{i,c_i} - {}^{i,p_i}\ddot{r}_{i,p_i}). \tag{4.105}
$$

All the quantities except for f_{Ck} $(k \in \mathcal{E}_C)$ used to compute $\ddot{q}_i$ are already computed in the assembly phase.

The computations for disassembling subchain C by removing joint i consist of following three steps:

(1) compute f_{Ci} (Eq.(4.103)),
(2) compute $\ddot{r}_{i,X}$ $(X = A, B)$ (Eq.(4.104)), and
(3) compute $\ddot{q}_i$ (Eq.(4.105)).

4.5.4 Closed Kinematic Chains

Including closed loops does not require significant modifications of the algorithm, except that a joint may be connecting two links in the same subchain. In this case, however, $\Gamma_{i,i}$ may not be invertible. This problem happens when the constraint forces are indeterminate, i.e. the structure is redundantly constrained or when the constraints are inconsistent. Fig. 4.6 illustrates the first case. In Fig. 4.6, the motion of link B in the x axis is redundantly constrained because the motion is already constrained by the constraint on link A. Therefore we cannot compute the individual constraint forces in the x direction; only their sum is defined.

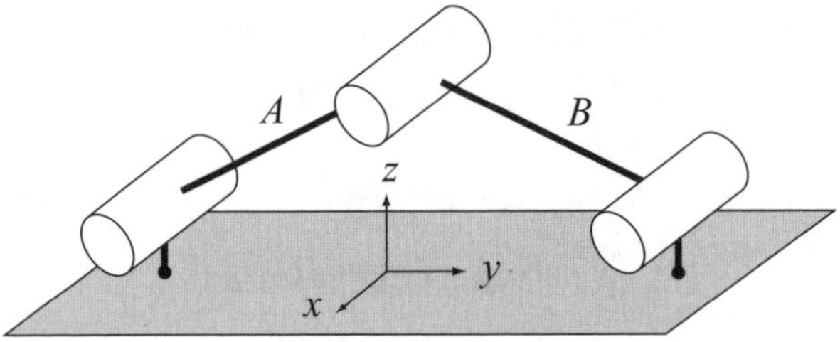

Fig. 4.6. A situation that results in indeterminate constraint forces.

4.6 Parallel Computation

4.6.1 Scheduling

The method described in the previous section assumes nothing about the order of adding and removing joints in the assembly and disassembly phases. In fact, it is the scheduling that determines the parallelism and total computational cost. In general, if we try to increase the parallelism the total computational cost grows up accordingly and vise versa; therefore, it is less efficient to apply a schedule intended for larger number of processes than available. One of the advantages of our approach is that we can customize the parallelism and the total computational cost only by changing the schedule.

Figs. 4.7 and 4.8 show two possible schedules for assembling an 8-link serial chain. It is obvious that the former one has higher parallelism, since it allows four processes to run in parallel at the first step and requires only three steps in total. The latter one, on the other hand, allows only two parallel processes and requires four steps in total. If we have more than four processors, therefore, it is better to apply the former schedule. Otherwise it is better to use the latter order because its total computational cost is worse than the latter.

The issue here is how to determine the schedule of the assembly and disassembly phases that makes the best use of the available processors. A qualitative strategy is shown to obtain the optimal scheduling for a given kinematic chain and the number of processors. It is difficult to give an algorithmic strategy for general cases and would be included in future works.

A schedule can be expressed by a binary tree like (a)–(c) in Fig. 4.9. These binary trees represent different schedules for the kinematic chain in left-hand side of Fig. 4.9. Each node represents a subchain and is labeled by the number of the last joint added to assemble the subchain. Two edges starting from a node point the two subchains connected by the joint. Null-pointing edges mean that the corresponding subchain consists of a single link.

A binary tree gives an intuitive idea of the parallelism and efficiency of the schedule: the number of leaf nodes indicates the parallelism, while the depth is almost proportional to the computation time when there exists more processors than the

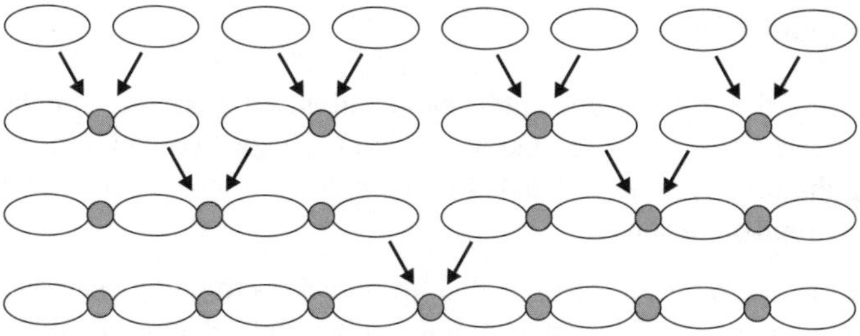

Fig. 4.7. A schedule for an 8-link serial chain.

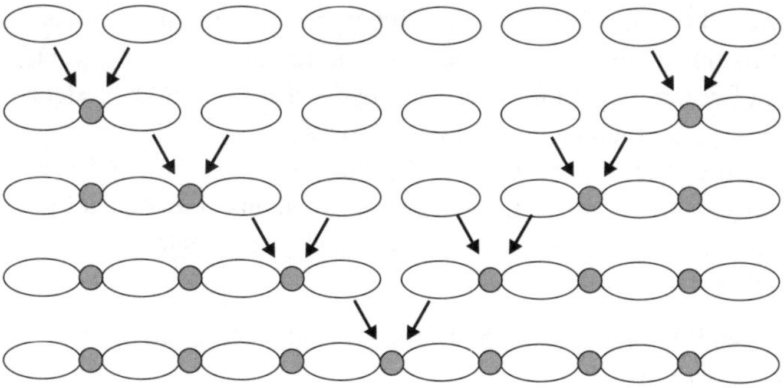

Fig. 4.8. Another possible schedule.

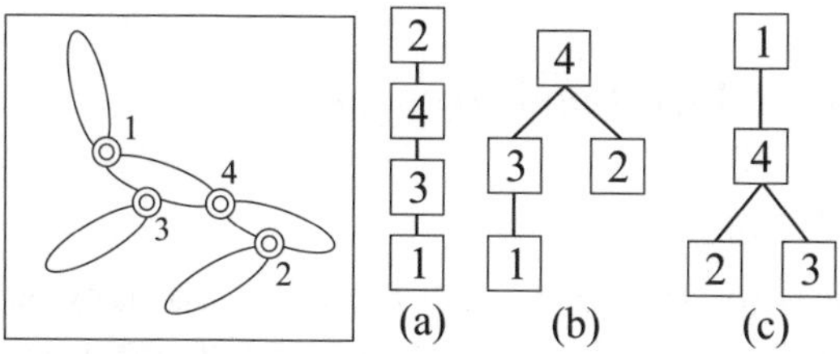

Fig. 4.9. Binary tree representation of different schedules for a kinematic chain.

number of the leaf nodes. The properties derive two strategies to build a binary tree for optimal scheduling:

(1) The number of leaf nodes should be close to the number of the processors.
(2) The depth of leaf nodes should be uniform.

The first strategy ensures that the schedule is not "over-parallelized" because the total computational cost grows as the parallelism increases. The second strategy tries to distribute the load evenly to keep the idle time of the processors as small as possible.

4.6.2 Communication

Before starting the forward dynamics computation, all processes should know the current state, which is either the initial condition for the simulation or the result of integrating the accelerations computed in the previous step. There are two possible ways to accomplish this task:

- Gather the accelerations computed at all processes to one process and integrate them there; then broadcast the new joint values and velocities to all processes.
- Distribute the joint accelerations to all processes and integrate them independently at each process.

The first method requires less number of inter-process communications, so it is probably better for most cases. The second method requires more communications, but the number of communications and integrated variables could be reduced by removing unnecessary communications for each process.

In addition to the final result, we need to pass some internal variables during the forward dynamics computation. Suppose subchains A and C in Fig. 4.5 are processed in different processes p_a and p_c. The following values should be passed between the two processes:

(1) In the assembly phase, send the following data from p_a to p_c:
  a) $\boldsymbol{\Lambda}_{Am,n}(m, n \in \mathcal{E}_A)$
  b) ${}^{A}\ddot{\boldsymbol{r}}_{m,A}(m \in \mathcal{E}_A)$
(2) In the disassembly phase, send $\boldsymbol{f}_i$ from p_c to p_a.

4.7 Computational Cost and Accuracy

4.7.1 Complexity

Computational costs for $\boldsymbol{P}_{i,i}, \boldsymbol{\Gamma}_{i,i}^{-1}$ and ${}^{C}\boldsymbol{f}_{Ci}$ in the assembly phase and $\ddot{\boldsymbol{q}}_i$ in the disassembly phase are bounded in all joints for arbitrary open and closed kinematic chains. Computing other quantities, on the other hand, requires $O(n_C)$ or $O(n_C^2)$ computation where n_C is the number of joints in $\mathcal{E}_C$. Since n_C can be any number for complicated kinematic chains, the total asymptotic complexity may be larger than $O(N)$.

However, we can achieve $O(N)$ complexity by giving upper bound to n_C. In all open kinematic chains, for example, n_C can always be reduced to no more than three by splitting links with more than three neighboring joints and inserting fixed joints between the pieces, as shown in Fig. 4.10. The number of neighboring joints of every subchain is limited to three when appropriate scheduling is set. Link splitting increases the number of joints, thus the total computational cost also increases. However, the number never exceeds twice of the original because the increase is always less than the number of joints in the original chain. Therefore, the total asymptotic complexity is still $O(N)$.

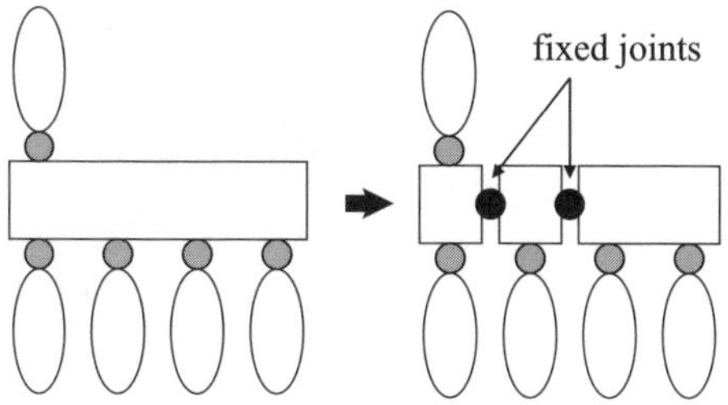

Fig. 4.10. Link splitting.

It is possible to apply link splitting to closed kinematic chains to reduce n_C. It is difficult, however, to prove that link splitting always gives upper bound for n_C on general closed kinematic chains. We cannot guarantee that the proposed algorithm achieves $O(N)$ complexity for general closed kinematic chains, although it does for most of the practical ones.

4.7.2 Improving the Performance

There are several ways to reduce the number of arithmetic operations. For example, $\boldsymbol{M}_k$ and $\boldsymbol{J}_{i,k}$ can be computed before running the simulation and do not affect the runtime efficiency because they remain constant for any configuration, thanks to the usage of local coordinates. $\boldsymbol{M}_k^{-1}$, $\boldsymbol{H}_{Ci,p_i}$ and $\boldsymbol{H}_{Ji,p_i}$, therefore, are also computed before the simulation. Even for variables that should be computed for each step, precomputing frequently-used components is also a common technique.

Another way is to take the advantage of the properties of $\boldsymbol{K}_{Ci}$ and $\boldsymbol{K}_{Ji}$ that they consist of only 0 or 1 in most of the practical joint types. In such cases, we are essentially only picking up some of the rows or columns of a matrix or a vector by multiplying these matrices. We can therefore eliminate the multiplications and additions involved in those matrix-matrix or matrix-vector multiplications.

The most costly processes of the algorithm are Eqs.(4.85) and (4.102), which involve the inverse of $\boldsymbol{\Gamma}_{i,i}$. In general, it is a good idea to utilize a library for linear algebra such as LAPACK [2] which provides efficient and stable functions for solving linear equations. We can reduce the computational cost by taking advantage of the fact that $\boldsymbol{\Gamma}_{i,i}$ is always symmetric and positive definite. In LAPACK, for example, a special solver designed for symmetric positive-definite coefficient matrix called dposv() saves much computation compared to the generic solver dgesvx(). In our experiment, the implementation using dposv() took only two-thirds of the time spent by that using dgesvx().

It is also better to use multi-DOF joints rather than dividing them into single-DOF joints in terms of both the size of $\boldsymbol{\Gamma}_{i,i}$ and the number of links. If a chain contains a spherical joint, for example, it is more efficient to model it as a spherical joint rather than dividing it into three rotational joints.

The whole algorithm requires approximately 820 multiplications and 690 additions per link for serial kinematic chains if all joints have 1DOF, including the

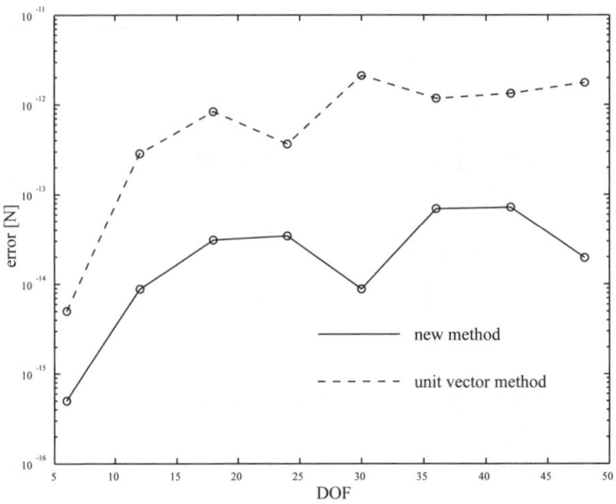

Fig. 4.11. Comparison of errors in the assembly-disassembly algorithm and unit vector method.

forward kinematics to compute the positions, orientations, and velocities of the links (assuming that Gaussian Elimination is used for matrix inversion). If all links are connected by 3DOF joints, the cost is reduced to 460 multiplications and 430 additions per link.

4.7.3 Accuracy

Matrix inversion is one of the major sources of numerical errors. The assembly-disassembly algorithm has an advantage over the algorithm described in Chapter 2 in this sense because the size of the matrices to be inverted are smaller than 6×6, whereas the previous algorithm involves inversion of the large joint-space mass matrix.

Fig. 4.11 compares the accuracy of the assembly-disassembly algorithm with that of unit vector method [96]. The accuracy test was performed as follows:

(1) Using the forward dynamics algorithm, compute the joint acceleration for zero joint torques.
(2) Using Newton-Euler formulation [71] for inverse dynamics, compute the joint torques to generate the acceleration computed in (1), which are ideally zero.
(3) Use the length of the torque vector computed in (2) as the measure of the error.

The error in step (2) should be much smaller than that in step (1) because Newton-Euler formulation does not contain divisions. The tests were executed for serial chains with 6 to 48DOF. The assembly-disassembly algorithm was more accurate than unit vector method for all cases.

4.8 Relationship with Other Algorithms

The first algorithm to achieve $O(\log N)$ asymptotic complexity was Constraint Force Algorithm (CFA) [26]. This algorithm used the orthogonality of $\boldsymbol{K}_{Ci}$ and $\boldsymbol{K}_{Ji}$ to factorize the inverse mass matrix $\boldsymbol{S}_X$. The original version was only applicable to serial chains, but later extended to allow short branch from the main chain [25].

Featherstone [23, 24] also developed an $O(\log N)$ algorithm called Divide-and-Conquer Algorithm (DCA) based on his earlier work. DCA uses the articulated body inertia [22] of the subchains to compute the forward dynamics.

The assembly-disassembly algorithm has close relationships with both algorithms. It is similar to CFA in the sense that the constraint forces are also computed during the forward dynamics computation. However, we do not assume the orthogonality of the constraint and motion space, while CFA does. In addition, the $O(\log N)$ complexity of our algorithm is valid for all open kinematic chains and most of the practical closed kinematic chains. Similarity with DCA is the concept of assembling and disassembling the target chain for parallel processing. The underlying computations are, however, very different in that the constraint forces are computed as well as the joint accelerations in our algorithm. This feature becomes an advantage for some applications that requires the information of constraint forces because they are computed directly without additional inverse dynamics step. For example, the constraint force is essential for contact models like the one to be presented in Chapter 5. We can also make sure that the mechanism does not collapse due to large constraint force at the joints.

4.9 Simulation Examples

The presented algorithm was implemented on a cluster of 8 workstations, each node with a PentiumIII 1GHz processor. The nodes are connected by Myrinet and have parallel computation environment SCore [73] installed.

4.9.1 Computation Time

Table 4.1 shows the computation time for serial kinematic chains of 8 to 32 links connected by 3DOF spherical joints. The links are distributed to all processes evenly, except for the last case (marked * in the table), where we used 4 processes but the numbers of links assigned to them are not even. When a serial chain is divided into 4 subchains processed by 4 processes, the middle two subchains have two joints in $\mathcal{E}$, while those at the end have only one. Therefore, the computational costs of the processes handling the middle two subchains are slightly larger than the other two even if the numbers of the assigned links are the same. In the last case we moved one link each from the middle subchains to the other two and tried to make the computational costs, rather than the number of links, of the processes even. As a result, the computation time was reduced greatly from the case of 4 processes each with the same number of links.

Table 4.1. Computation time for serial chains (ms).

links	8	16	32
1 procs	1.31	2.75	6.08
2 procs	0.984	1.87	3.93
4 procs	0.897	1.70	3.39
8 procs	—	1.57	2.90
4 procs*	—	1.58	3.16

Table 4.2. Computation time for human figures (ms).

DOF	34	48
1 procs	3.66	4.85
2 procs	2.49	2.93
4 procs	2.22	2.49

Table 4.2 shows the computation time for free-flying human figures of 34 and 48DOF on 1 to 4 processes. Since typical human figures contain four limbs, it is natural to divide the chain into four segments and assign one process each. Using more than four processes is generally meaningless for human figures.

4.9.2 Dynamics Simulation of Human Figures

Fig. 4.12 shows snapshots from dynamics simulation of a human figure including structure changes. The figure initially held the environment by the both hands, and then released the right and left hands. The joints are controlled by a PD controller with the fixed reference joint angles. The computation time in serial computation varies from 3.4 to 5.0 ms depending on the number of connections between the hands and the environment. Although the model contains a closed loop when both hands are grasping the bar, we do not suffer from indeterminate constraint force because the arms have enough DOF to make the constraints independent.

4.10 Summary

This chapter presented a new forward dynamics algorithm which has $O(N)$ asymptotic complexity for serial computation and $O(\log N)$ complexity for parallel computation on $O(N)$ processors. The contributions of the chapter are summarized by the following four points:

(1) An $O(N)$ formulation of forward dynamics algorithm for open and closed kinematic chains was developed. The complexity is reduced even more to $O(\log N)$ by applying parallel computation. The formulation consists of assembly and disassembly phases. The $O(N)$ complexity is achieved by focusing on the relationships between the force and acceleration at the end links of each subchain.

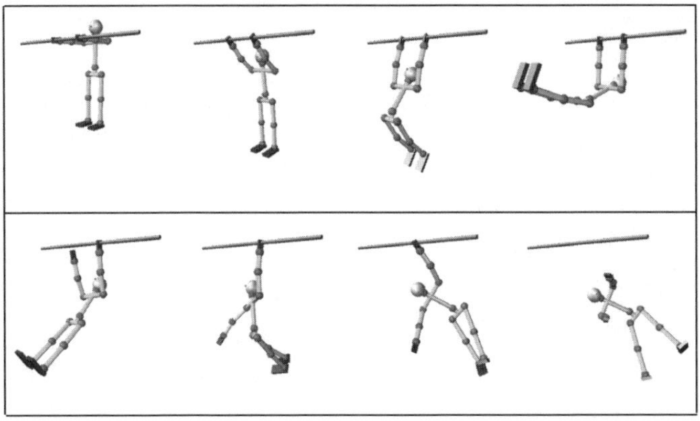

Fig. 4.12. Dynamics simulation of a structure-varying human figure.

(2) It was shown that the parallelism of the proposed algorithm can be tuned easily for any kinematic chains on any number of processors by optimizing the order of adding and removing joints in the assembly and disassembly phases.
(3) An qualitative idea for optimizing the schedule for parallel computation was shown using the binary tree representation of scheduling. The parallelism and total computation time are characterized by the width and depth of the tree respectively.
(4) Simulation examples demonstrated the effectiveness of the new algorithm and parallel computation.

5

Iterative Computation of Contact Force for Rigid Collision/Contact Model

5.1 Introduction

This chapter extends the method described in Chapter 2 to handle collisions and contacts. The model yields relatively stable results with large time steps because it is an analytical approach. It also reduces the computation for optimization by applying an iterative trial-and-error procedure to find the constraints and contact forces that satisfy the unilateral conditions while most analytical methods apply numerical optimization techniques. The equations also serve as the basis of the *dynamics filter* presented in Chapter 7.

The advantage of analytical approaches to collision and contact modeling is the stability of the result. Since the constraint conditions are explicitly specified, it is unlikely that we encounter unrealistic constraint forces which occasionally occurs in penalty-based approaches, as long as the optimization problem has a solution. However, the problem is that the optimization process is usually too time-consuming for realtime simulation.

This chapter includes two technical improvements of the method presented in Chapter 2. Firstly, we enable the computation of the constraint forces by connecting the virtual links by 6-DOF free joints. We can set arbitrary constraints to the joint values of the free joint and compute the constraint force to maintain the constraint. It also turns out that we do not need to recompute the degrees of freedom and reselect the generalized coordinates when the constraint condition has been changed.

Secondly, an iterative procedure is developed to find the constraint condition and constraint forces that satisfy the unilateral constraints. It repeats the process of checking whether the constraint forces for a set of constraint conditions satisfy the unilateral conditions and, if not, modify the constraint conditions and compute the constraint forces again. Since the computational cost of each iteration is very small, the whole procedure is more efficient than applying optimization process.

This chapter was adapted from, by permission, K. Yamane and Y. Nakamura, "Dynamics Filter—Concept and Implementation of On-Line Motion Generator for Human Figures," IEEE Transactions on Robotics and Automation, vol.19, no.3, pp.421–432, 2003.

This chapter is organized as follows. First, we review the existing techniques for modeling collisions and contacts in Section 5.2. In Section 5.3, the equation of motion of Chapter 2 is extended to compute the constraint forces for a known constraint condition. Then the overview of the iterative procedure is shown in Section 5.4, followed by the details in Section 5.5. Section 5.6 deals with the collision model that computes the discontinuous change of joint velocities on collision. Finally, several simulation examples are presented in Section 5.7.

5.2 Related Work

Simulation of collisions and contacts has been discussed for many years and a number of methods have been proposed. They are basically divided into two categories: penalty-based methods and analytical methods.

In penalty-based methods ([55, 38]), contact forces are generated by virtual springs and dampers at the contact points. These approaches are commonly used in commercial software packages for dynamics simulation because of the easy implementation. However, the problem is that this approach requires extremely precise and time-consuming simulation due to the discrete integration of stiff system. It is also difficult to find parameters that yield realistic results.

Analytical methods compute the contact forces that satisfy the unilateral conditions using optimization techniques such as quadratic programming (QP) [50], Linear Complementarity Problem (LCP) solvers [78, 91], or other simplified techniques [10]. These methods can produce relatively stable results with large sampling time but solving optimization problems tends to be time-consuming and requires a simplification of the problem. There are some other approaches such as impulse-based method [59] which is powerful in systems where bouncing occurs more frequently than stick contacts.

As a whole, efficient and precise simulation of collisions and contacts still remains an open research issue. One approach to overcome this problem is to make use of the nature of collisions and contacts in question to simplify the problem [10]. We develop an efficient method for collision/contact simulation taking advantage of the observation that most contacts in human motion do not produce bouncing. Our method is basically an analytical approach but, instead of the mathematical optimization techniques, we apply a trial-and-error process to find the constraint condition and contact forces that satisfy the unilateral conditions. Collisions are modeled as discontinuous change of the joint velocities to enforce zero relative velocity between the contact pairs and to avoid impulsive accelerations on impact.

5.3 Basic Equations

Suppose a link in a robot *Link R* is in contact with an environment link *Link E*. Following the strategy described in Chapter 3, we create a virtual link of *Link R*, named *Link Rv*, as a child link of *Link E* as illustrated in Fig. 5.1.

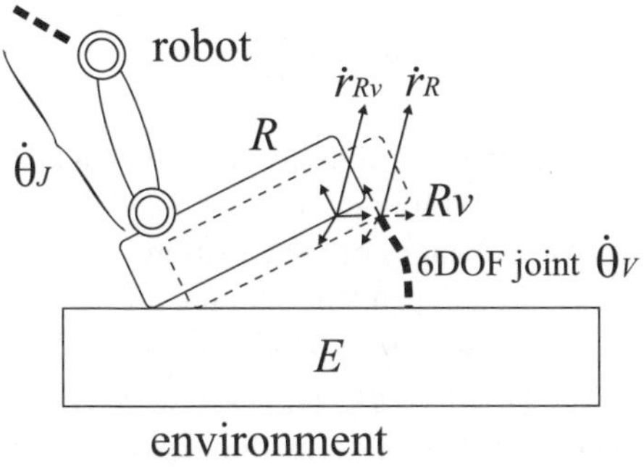

Fig. 5.1. Virtual link for the contact model.

In a bilateral constraint, *Link Rv* would be connected to *Link E* through a joint of a known type. In a contact constraint, on the other hand, we do not know the constraint condition at this stage. In the extended method, we place a 6-degrees-of-freedom (DOF) joint between *Link Rv* and *Link E* and impose an additional constraint on the joint velocity of the 6-DOF joint.

Let N_{DOF} denote the total degrees of freedom of the robot including the 6 DOF of the root link, $\dot{\boldsymbol{\theta}}_J \in \boldsymbol{R}^{N_{DOF}}$ the vector composed of joint velocities of the robot, $\dot{\boldsymbol{\theta}}_V \in \boldsymbol{R}^6$ the joint velocity of the 6-DOF joint between *Link Rv* and *Link E*, and $\dot{\boldsymbol{r}}_R \in \boldsymbol{R}^6$ and $\dot{\boldsymbol{r}}_{Rv} \in \boldsymbol{R}^6$ the vectors composed of the linear and angular velocities of *Link R* and *Link Rv* respectively. Again following our strategy, we virtually cut the joint between *Link Rv* and *Link E* to obtain a virtual open kinematic chain. The two sets of joint velocities, $\dot{\boldsymbol{\theta}}_A$ including those of actuated joints and $\dot{\boldsymbol{\theta}}_O$ including those in the virtual open kinematic chain, are written as follows respectively:

$$\dot{\boldsymbol{\theta}}_O = \dot{\boldsymbol{\theta}}_J \tag{5.1}$$

$$\dot{\boldsymbol{\theta}}_A = \begin{pmatrix} \dot{\boldsymbol{\theta}}_V \\ \dot{\boldsymbol{\theta}}_J \end{pmatrix} \tag{5.2}$$

and the generalized coordinates $\boldsymbol{\theta}_G$ are yet to be selected.

$\dot{\boldsymbol{r}}_R$ and $\dot{\boldsymbol{r}}_{Rv}$ are described as

$$\dot{\boldsymbol{r}}_R = \boldsymbol{J}_R \dot{\boldsymbol{\theta}}_J \tag{5.3}$$

$$\dot{\boldsymbol{r}}_{Rv} = \boldsymbol{J}_V \dot{\boldsymbol{\theta}}_V \tag{5.4}$$

where $\boldsymbol{J}_R \in \boldsymbol{R}^{6 \times N_{DOF}}$ and $\boldsymbol{J}_V \in \boldsymbol{R}^{6 \times 6}$ are the Jacobian matrices of $\boldsymbol{r}_R$ with respect to $\boldsymbol{\theta}_J$ and $\boldsymbol{r}_{Rv}$ with respect to $\boldsymbol{\theta}_V$, respectively, that is,

$$\boldsymbol{J}_R = \frac{\partial \boldsymbol{r}_R}{\partial \boldsymbol{\theta}_J} \tag{5.5}$$

$$\boldsymbol{J}_V = \frac{\partial \boldsymbol{r}_{Rv}}{\partial \boldsymbol{\theta}_V}. \tag{5.6}$$

Since $\dot{\boldsymbol{r}}_{Rv}$ should be equal to $\dot{\boldsymbol{r}}_R$, the following equation holds:

$$\begin{pmatrix} \boldsymbol{J}_V & -\boldsymbol{J}_R \end{pmatrix} \begin{pmatrix} \dot{\boldsymbol{\theta}}_V \\ \dot{\boldsymbol{\theta}}_J \end{pmatrix} = \boldsymbol{O}. \tag{5.7}$$

This equation corresponds to the closed loop constraint equation (2.10) in Chapter 2. We consider single contact case for simplicity, although handling multiple contacts requires only straightforward changes of the equations.

Next step is to extract six independent columns of the coefficient matrix of the left-hand side of Eq.(5.7) to select the generalized coordinates $\boldsymbol{\theta}_G$ and compute the two Jacobian matrices $\boldsymbol{W} \triangleq \partial\boldsymbol{\theta}_O/\partial\boldsymbol{\theta}_G$ and $\boldsymbol{S} \triangleq \partial\boldsymbol{\theta}_A/\partial\boldsymbol{\theta}_G$. By the definition of $\boldsymbol{J}_V$, it is usually just a coordinate transformation from $\dot{\boldsymbol{\theta}}_V$ to $\dot{\boldsymbol{r}}_{Rv}$ because they represent the same physical values—linear and angular velocity of *Link Rv*—in different frames. Therefore, the columns of $\boldsymbol{J}_V$ is obviously independent of each others, which implies that $\boldsymbol{\theta}_J$ can be the generalized coordinates of the system, and that the Jacobian matrix of dependent joints (in this case $\boldsymbol{\theta}_V$) with respect to the generalized coordinates are computed by

$$\boldsymbol{H} = -\boldsymbol{J}_V^{-1}(-\boldsymbol{J}_R) = \boldsymbol{J}_V^{-1}\boldsymbol{J}_R \tag{5.8}$$

as described in Eq.(2.17). Therefore, $\boldsymbol{W}$ and $\boldsymbol{S}$ are formed as

$$\boldsymbol{W} = \boldsymbol{I} \tag{5.9}$$

$$\boldsymbol{S} = \begin{pmatrix} \boldsymbol{H} \\ \boldsymbol{I} \end{pmatrix} \tag{5.10}$$

where $\boldsymbol{I}$ is the identity matrix of the appropriate size.

The fact that the joint values of the robot can be used as the generalized coordinates indicates that we do not have to reselect the generalized coordinates and recompute the inertial matrix even if the constraint condition or the number of contacts changes. In addition, we can easily compute the inertial matrix in the generalized coordinate space because it coincides with the joint space inertial matrix of the robot.

Now we define some variables related to force and inertia. Let $\boldsymbol{A} \in \boldsymbol{R}^{N_{DOF} \times N_{DOF}}$ denote the inertial matrix in the generalized coordinate space and $\boldsymbol{b}$ denote the joint torques required to produce zero accelerations on the robot without contact, both of which can be computed by applying well-known methods such as [96] to the robot. Also let $\boldsymbol{\tau}_J \in \boldsymbol{R}^{N_{DOF}}$ denote the vector composed of the joint torques of the robot and $\boldsymbol{\tau}_V \in \boldsymbol{R}^6$ the joint force and torque of the 6-DOF joint between *Link E* and *Link Rv*. $\boldsymbol{\tau}_V$ is actually identical to the constraint force and moment. Note that the elements of $\boldsymbol{\tau}_J$ corresponding to the 6 DOF of the root link are always zero. To clarify this point, we introduce another vector $\boldsymbol{\tau}_A \in \boldsymbol{R}^{N_{DOF}-6}$ that includes only the torques of the actuated joints, and a matrix $\boldsymbol{H}_J \in \boldsymbol{R}^{N_{DOF} \times (N_{DOF}-6)}$ that maps $\boldsymbol{\tau}_A$ to $\boldsymbol{\tau}_J$ as

$$\boldsymbol{\tau}_J = \boldsymbol{H}_J^T \boldsymbol{\tau}_A. \tag{5.11}$$

Using these notations, the generalized force acting to the system $\boldsymbol{\tau}_G \in \boldsymbol{R}^{N_J}$ is computed as

$$\begin{aligned}\boldsymbol{\tau}_G &= \boldsymbol{S}^T \begin{pmatrix} \boldsymbol{\tau}_V \\ \boldsymbol{\tau}_J \end{pmatrix} + \boldsymbol{f}_{ext} \\ &= \boldsymbol{H}^T \boldsymbol{\tau}_V + \boldsymbol{H}_J^T \boldsymbol{\tau}_A + \boldsymbol{f}_{ext} \end{aligned} \tag{5.12}$$

where $\boldsymbol{f}_{ext}$ includes the effect of all known external forces.

Noting that the generalized coordinates of the system are $\boldsymbol{\theta}_J$, the equation of the motion becomes [63]

$$\boldsymbol{A}\ddot{\boldsymbol{\theta}}_J + \boldsymbol{b} = \boldsymbol{\tau}_G. \tag{5.13}$$

On the other hand, the relationship of $\ddot{\boldsymbol{\theta}}_J$ and $\ddot{\boldsymbol{\theta}}_V$ is described as

$$\ddot{\boldsymbol{\theta}}_V = \boldsymbol{H}\ddot{\boldsymbol{\theta}}_J + \dot{\boldsymbol{H}}\dot{\boldsymbol{\theta}}_J. \tag{5.14}$$

If *Link R* is fixed to *Link E*, we know that $\ddot{\boldsymbol{\theta}}_V = \boldsymbol{O}$, so from Eqs.(5.12)–(5.14) we obtain the following linear equation with unknowns $\ddot{\boldsymbol{\theta}}_J$ and $\boldsymbol{\tau}_V$:

$$\begin{pmatrix} \boldsymbol{A} & -\boldsymbol{H}^T \\ \boldsymbol{H} & \boldsymbol{O} \end{pmatrix} \begin{pmatrix} \ddot{\boldsymbol{\theta}}_J \\ \boldsymbol{\tau}_V \end{pmatrix} = \begin{pmatrix} \boldsymbol{H}_J^T \boldsymbol{\tau}_J + \boldsymbol{f}_{ext} - \boldsymbol{b} \\ -\dot{\boldsymbol{H}}\dot{\boldsymbol{\theta}}_J \end{pmatrix}. \tag{5.15}$$

The coefficient matrix of the left-hand side of this equation is always square, and invertible if $\boldsymbol{J}_R$ has full row rank, in which case we can compute the joint accelerations $\ddot{\boldsymbol{\theta}}_J$ as well as the constraint force $\boldsymbol{\tau}_V$ by solving Eq.(5.15).

This equation is identical to the equation of motion of constrained systems found in many literatures [12, 51], except that the generalized coordinates and inertial matrix are in joint space. Also, we encounter a matrix $\boldsymbol{H}\boldsymbol{A}^{-1}\boldsymbol{H}^T$ while computing the solution of Eq.(5.15), which is identical to the operational space inertia matrix [48] in single contact case or the extended operational space inertia matrix [15] if we have multiple contacts.

The same scheme can be applied to the computation of discontinuous change of joint velocities on link connection described in Section 3.4, yielding a method to compute the impact force on connection. This is used in the simulation of collisions described in Section 5.6.

If *Link R* and *Link E* are connected by contact constraint, the constraint condition varies depending on the contact state. Suppose the constraints for the acceleration $\ddot{\boldsymbol{\theta}}_V$ are described by n_C $(0 \leq n_C \leq 6)$ equations

$$\boldsymbol{K}_C \ddot{\boldsymbol{\theta}}_V = \boldsymbol{O}, \tag{5.16}$$

where $\boldsymbol{K}_C$ is an $(n_C \times 6)$ matrix. Using this notation, Eq.(5.15) is modified as

$$\begin{pmatrix} \boldsymbol{A} & -\boldsymbol{H}_C^T \\ \boldsymbol{H}_C & \boldsymbol{O} \end{pmatrix} \begin{pmatrix} \ddot{\boldsymbol{\theta}}_J \\ \boldsymbol{\tau}_C \end{pmatrix} = \begin{pmatrix} \boldsymbol{t} \\ \boldsymbol{a} \end{pmatrix} \tag{5.17}$$

where

$$\begin{aligned} \boldsymbol{H}_C &\triangleq \boldsymbol{K}_C \boldsymbol{H} \\ \boldsymbol{\tau}_C &\triangleq \boldsymbol{K}_C^T \boldsymbol{\tau}_V \\ \boldsymbol{t} &\triangleq \boldsymbol{H}_J^T \boldsymbol{\tau}_J + \boldsymbol{f}_{ext} - \boldsymbol{b} \\ \boldsymbol{a} &\triangleq -\dot{\boldsymbol{H}}_C \dot{\boldsymbol{\theta}}_J. \end{aligned}$$

Eq.(5.17) is the general equation of motion of a human figure subject to constraints with the environment and serves as the basic equation for collision and contact simulation.

5.4 Overview

At this stage, we can compute the contact force for a known constraint condition by solving Eq.(5.17). In this section, we present a method for finding the constraint condition and the constraint force that satisfy the unilateral conditions, assuming that the normal relative velocity is zero. If the links come into contact with nonzero normal relative velocity, it is set to zero by the collision computation described in Section 5.6.

The idea of the trial-and-error procedure is to compute the constraint forces and moments that satisfy the temporarily assumed constraints, and then check if they satisfy the unilateral conditions. If they do not, we change the constraint condition and compute the constraint force again. The process is repeated until all the constraint forces and moments satisfy the unilateral conditions.

The whole procedure is summarized as follows:

(1) Initialize the process by setting all possible constraints.
(2) Compute the constraint forces and moments by Eq.(5.17).
(3) Check if the constraint forces and moments satisfy the unilateral conditions as described in Section 5.5.3.
(4) If invalid constraint forces were found, modify the constraint as described in Section 5.5.4 and return to (2), otherwise proceed to (5).
(5) Compute the joint accelerations using Eq.(5.17).

5.5 Details

5.5.1 Contact Coordinate

For simple representation of various constraints, we place the virtual link frame, or the contact coordinate frame, as illustrated in Fig. 5.2 at each contact pair. The z axis is parallel to the normal vector of the contact surface and the other axes and the position are set as follows depending on the contact state:

- Point contact: position is set to the contact point, and the direction of x and y axes are arbitrary.
- Line contact: the x axis is parallel to the contact line, and the position can be any point on the line.
- Face contact: the position can be any point in the contact area, and the directions of x and y axes are arbitrary.

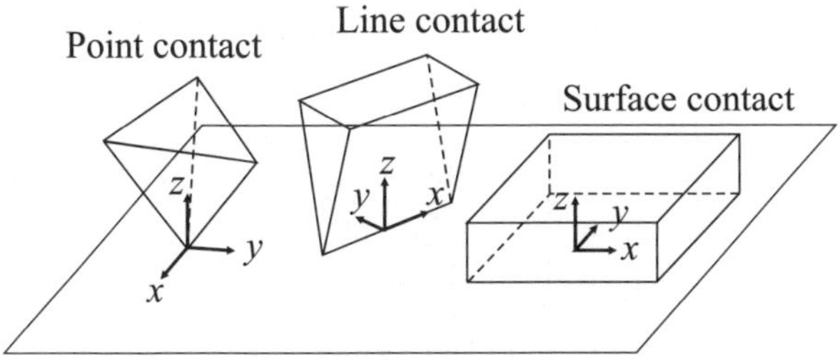

Fig. 5.2. Contact coordinates for each contact state.

By virtue of the definition of the contact coordinate, the constraint matrix $\boldsymbol{K}_C$ has a very simple structure composed of only 1's and 0's for all practical constraint conditions. For example, if a pair of links in line contact is fully constrained, $\boldsymbol{K}_C$ becomes

$$\boldsymbol{K}_C = \begin{pmatrix} 1\,0\,0\,0\,0\,0 \\ 0\,1\,0\,0\,0\,0 \\ 0\,0\,1\,0\,0\,0 \\ 0\,0\,0\,0\,1\,0 \\ 0\,0\,0\,0\,0\,1 \end{pmatrix} \tag{5.18}$$

because only the rotation around x axis is unconstrained. If the links are slipping, $\boldsymbol{K}_C$ would be

$$\boldsymbol{K}_C = \begin{pmatrix} 0\,0\,1\,0\,0\,0 \\ 0\,0\,0\,0\,1\,0 \\ 0\,0\,0\,0\,0\,1 \end{pmatrix} \tag{5.19}$$

because the motion in x and y axes are not constrained either.

In the following discussion, the spatial velocity of the contact coordinate and the constraint force/moment are expressed in the contact coordinate and denoted by vectors $(v_x\ v_y\ v_z\ \omega_x\ \omega_y\ \omega_z)^T$ and $(f_x\ f_y\ f_z\ n_x\ n_y\ n_z)^T$ respectively.

5.5.2 List of Constraint Conditions

Table 5.1 lists all the possible constraint conditions (sets of constrained directions) C_0–C_7 and the feasibility of the constraint conditions at each contact state. The column *constraints* shows whether each direction is constrained (◯) or not ($\times$) in each constraint condition, while the column *state* shows whether the constraint condition is feasible (◯) or not ($\times$) at each contact state. In point contact, for example, only constraint conditions C_3 (position constrained), C_6 (only normal direction constrained) and C_7 (no constraint) are feasible because point contacts cannot produce any moment.

Conditions C_1 and C_5, where ω_y is constrained but ω_x is not, are possible only at line contacts. Similarly, condition C_3, where v_x and v_y are constrained but ω_z is not, is possible at point contacts. The reason is that in face and line contacts, if ω_z is not

Table 5.1. Possible constraint conditions for each contact state.

ID	constraints						state		
	v_x	v_y	v_z	ω_x	ω_y	ω_z	face	line	point
C_0	○	○	○	○	○	○	○	×	×
C_1	○	○	○	×	○	○	×	○	×
C_2	○	○	○	×	×	○	○	○	×
C_3	○	○	○	×	×	×	×	×	○
C_4	×	×	○	○	○	×	○	×	×
C_5	×	×	○	×	○	×	×	○	×
C_6	×	×	○	×	×	×	○	○	○
C_7	×	×	×	×	×	×	○	○	○

constrained, namely if the links are rotating around the normal vector, then the total friction force is determined uniquely because all points in the contact area, except for the center of rotation, are slipping. Although we may also consider the case where the slipping of the center of rotation is constrained, it is omitted for simplicity.

Some of the constraints are not feasible if the current relative velocity satisfies the following conditions:

- If $v_x^2 + v_y^2 > 0$ or $\omega_z \neq 0$, directions in v_x, v_y and ω_z are not constrained. That is, if the links are slipping, the relative motion in the tangential plane is not constrained.
- If $\omega_x^2 + \omega_y^2 > 0$, directions in ω_x and ω_y are not constrained.

5.5.3 Constraint Force Validity Check

After computing the constraint forces and moments, we execute the validity checks listed below. If any one of the checks fails, we immediately change the constraint condition as described in Section 5.5.4.

(1) Normal force
The normal force f_z should be in the repulsive direction, namely $f_z \geq 0$.

(2) Center of pressure (COP)
n_x and n_y should satisfy the condition that the COP, computed by $(n_y/f_z, -n_x/f_z, 0)$, is in the contact area. If not, the actual COP is set to the point in the contact area closest to the computed COP.

(3) Friction
According to Coulomb's friction law, the friction force should be smaller than the maximum static friction, namely,

$$\sqrt{f_x^2 + f_y^2} \leq \mu_S f_z \tag{5.20}$$

where μ_S is the static friction coefficient. If the friction force exceeds this limit, the links in contact start slipping in the direction computed by

$$\left(-\frac{f_x}{\sqrt{f_x^2+f_y^2}}\ -\frac{f_y}{\sqrt{f_x^2+f_y^2}}\ 0\right)^T. \tag{5.21}$$

(4) Twist moment
When rotation around z axis is constrained, n_z is the sum of the moments due to the static friction forces distributed over the contact area. Upper and lower limits of n_z depend on the distribution of the normal forces, and cannot be determined only by the total normal force f_z. Before dealing with this problem, we first focus on the twist moment around the COP, $\hat{n}_z$, when $v_x^2 + v_y^2 \neq 0$ or $\omega_z \neq 0$. Let us consider two extreme cases:
a) When the links are slipping and not spinning, namely $\sqrt{v_x^2 + v_y^2} \neq 0$ and $\omega_z = 0$, the friction forces are in the same direction and the net friction acts at the COP because each friction force is proportional to the normal force acting at each point. Therefore, the friction forces do not produce moment around the z axis at COP, so $\hat{n}_z = 0$.
b) When the links are spinning around the COP, namely $\sqrt{\hat{v}_x^2 + \hat{v}_y^2} = 0$ and $\omega_z \neq 0$, where $\hat{v}_x$ and $\hat{v}_y$ denote the slip velocity of COP, the friction forces act as concentric circles, so the torque around the z axis is maximized. Although it is impossible to compute the torque exactly, we can approximate it by

$$\hat{n}_z^{max} = -sgn(\omega_z) l \mu_D f_z \tag{5.22}$$

where l is the representative dimension of the contact area and μ_D is the dynamic friction coefficient, and $sgn(x)$ denotes the sign of x.

Taking the above observations into account, we approximate $\hat{n}_z$ by the following function:

$$\hat{n}_z = -sgn(\omega_z) p(r) l \mu_D f_z \tag{5.23}$$

where r is the distance between the COP and the center of rotation, $p(x)(x \geq 0)$ is a function which satisfies $p(0) = 0$ and $\lim_{x \to +\infty} = 1$ A similar discussion is used to estimate the maximum and minimum values of n_z when v_x, v_y and ω_z are constrained. In this case we have computed f_x, f_y and n_z required to constrain the motion in the three axes. If the friction force exceeds its limit described in Eq.(5.20) and the tangential directions have turned out to be unconstrained, we do not have to check the validity of n_z because the rotation around the z axis is not constrained either as discussed in Section 5.5.2. Otherwise, we first compute the ratio of the actual friction force with respect to the maximum friction force k_f by

$$k_f = \frac{\sqrt{f_x^2 + f_y^2}}{\mu_S f_z} \tag{5.24}$$

which is always between 0 and 1. Large k_f implies that the distributed friction forces are in almost the same directions. In the above discussion we observed that in slipping without spinning, where the directions of slipping frictions are exactly the same, we have $\hat{n}_z = 0$. Therefore we can say that if k_f is large, the absolute values of the limits for $\hat{n}_z$ will be small. Small k_f, on the other hand, implies the directions can vary, in which case $\hat{n}_z$ is maximized by applying static frictions tangential to concentric circles around the COP, and allows larger $\hat{n}_z$. Finally, the maximum absolute value of $\hat{n}_z$ can be estimated by

$$\hat{n}_z^{max} = sgn(\omega_z)(1 - k_f)l\mu_D f_z. \tag{5.25}$$

Note that $(1 - k_f)$ is used in place of $p(r)$ in Eq.(5.22). To check the validity of n_z, we first convert it to the moment around the COP by

$$\hat{n}_z = n_z - c_y f_x + c_x f_y \tag{5.26}$$

where $(c_x \; c_y \; 0)^T$ is the position of COP, and then check if $|\hat{n}_z| \leq \hat{n}_z^{max}$.

5.5.4 Transition among Constraint Conditions

Table 5.2 shows the transition among the constraint conditions when invalid contact forces were found. The rows show the current constraint conditions, whereas the columns show the new constraint condition for each check. The constraint condition numbers in the brackets are used in place of those out of the brackets for line contact because those constraint condition are not feasible in line contact. "—" indicates that the check is not executed because the corresponding direction is not constrained. For example, in condition C_4, which means that the links are slipping, the friction forces are not checked because the friction forces are computed from f_z using the dynamic friction model.

For a pair of the current constraint condition and the check failed, the new constraint condition candidates in the corresponding cell are sought from the left. The conditions previously visited or inadequate for the contact state are ignored, and the first available condition becomes the next constraint condition.

Note that, for example, negative f_z does not immediately lead to constraint condition C_7 (no constraint). In some situations, even if we had negative f_z for a certain constraint condition, we might get positive f_z after removing constraints in (v_x, v_y, n_z) or (ω_x, ω_y) directions.

Table 5.2. New constraint condition; the number of the check items coincides with the item number of the list in section 5.5.3.

from	check failed			
	(1)	(2)	(3)	(4)
$C_{0(1)}$	$C_{2,4(5),6,7}$	$C_{2,4(5),6,7}$	$C_{4(5),2,6,7}$	$C_{4(5),2,6,7}$
C_2	$C_{4(5),6,7}$	—	$C_{4(5),6,7}$	$C_{4(5),6,7}$
C_3	$C_{6,7}$	—	$C_{6,7}$	—
$C_{4(5)}$	$C_{2,6,7}$	$C_{2,6,7}$	—	—
C_6	C_7	—	—	—

The recomputation of the constraint forces requires the following three steps:

(1) Compute $\boldsymbol{H}_C$. Practically, the work needed here is just to select appropriate rows of $\boldsymbol{H}$.
(2) Compute $\boldsymbol{H}_C \boldsymbol{A}^{-1} \boldsymbol{H}_C^T$ and its inverse. This is not time consuming either because the matrix to be inverted is usually small and we do not have to recompute $\boldsymbol{A}$ and $\boldsymbol{A}^{-1}$.
(3) Compute $\boldsymbol{\tau}_C$.

5.5.5 Forces Subject to f_z

When the links are slipping, the v_x and v_y directions are not constrained and the friction forces are computed explicitly by

$$f_x = -s_x \mu_D f_z$$
$$f_y = -s_y \mu_D f_z$$

where $\boldsymbol{s} = (s_x \ s_y \ 0)^T$ is the unit vector in the slipping direction. We have to include the effect of friction forces into the equation of motion Eq.(5.17). However, f_z, on which the friction forces depend, is computed by solving this equation itself and f_x and f_y are not included in $\boldsymbol{\tau}_C$.

Let us denote the rows of $\boldsymbol{H}$ corresponding to x, y and z directions, namely the first, second and third rows, by $\boldsymbol{h}_x$, $\boldsymbol{h}_y$ and $\boldsymbol{h}_z$ respectively. Also denote the rest of $\boldsymbol{H}$ by $\boldsymbol{H}_n$ and the vector $(n_x \ n_y \ n_z)^T$ by $\boldsymbol{n}$. The contribution of the contact forces to the generalized force is computed by

$$\begin{aligned}
\boldsymbol{H}^T \boldsymbol{f} &= \left(\boldsymbol{h}_x^T \ \boldsymbol{h}_y^T \ \boldsymbol{h}_z^T \ \boldsymbol{H}_n^T \right) \begin{pmatrix} f_x \\ f_y \\ f_z \\ \boldsymbol{n} \end{pmatrix} \\
&= (-s_x \mu_D \boldsymbol{h}_x^T - s_y \mu_D \boldsymbol{h}_y^T + \boldsymbol{h}_z^T) f_z + \boldsymbol{H}_n^T \boldsymbol{n} \\
&= \left(-s_x \mu_D \boldsymbol{h}_x^T - s_y \mu_D \boldsymbol{h}_y^T + \boldsymbol{h}_z^T \ \boldsymbol{H}_n^T \right) \begin{pmatrix} f_z \\ \boldsymbol{n} \end{pmatrix} \\
&= \hat{\boldsymbol{H}}_C^T \boldsymbol{\tau}_C
\end{aligned} \tag{5.27}$$

where

$$\hat{\boldsymbol{H}}_C \triangleq \begin{pmatrix} -s_x \mu_D \boldsymbol{h}_x - s_y \mu_D \boldsymbol{h}_y + \boldsymbol{h}_z \\ \boldsymbol{H}_n \end{pmatrix}. \tag{5.28}$$

Eq.(5.27) means that adding the effect of dynamic friction to generalized force is equivalent to replacing $\boldsymbol{H}_C^T$ by $\hat{\boldsymbol{H}}_C^T$ in Eq.(5.17).

n_x, n_y and n_z are also handled in a similar way. When ω_x and ω_y directions are not constrained and the COP are known to be at $(p_x \ p_y \ 0)$ in the contact coordinate, n_x and n_y are computed by

$$n_x = p_y f_z \tag{5.29}$$
$$n_y = -p_x f_z. \tag{5.30}$$

When ω_z is not constrained n_z is computed by Eq.(5.23). Let $\boldsymbol{h}_{nu}(u = x, y, z)$ denote the rows of $\boldsymbol{H}$ corresponding to ω_u. If each direction is not constrained, the row of $\hat{\boldsymbol{H}}_C$ corresponding to v_z is increased by the following vectors:

$$\omega_x : p_y \boldsymbol{h}_{nx}$$
$$\omega_y : -p_x \boldsymbol{h}_{ny}$$
$$\omega_z : -sgn(\omega_z)p(r)l\mu_D \boldsymbol{h}_{nz}.$$

5.5.6 Limitations

The method has the following two limitations as the drawbacks of the simplifications:

- Multiple Contacts
  The problem of handling multiple contacts is the coupling between the contacts. We basically reduce the number of constraints if invalid constraint forces are found, and each contact pair never retries the constraint conditions that have been already checked. This scheme enables us to obtain the feasible constraint forces with $O(mn)$ computations, where m denotes the number of pairs in contact and n the number of possible contact conditions for each pair, while checking all possible sets of constraint conditions requires $O(n^m)$ computations. The problem is that, however, it might happen that the most desirable solution is never tried in the course of the trial-and-error process. This failure could lead to negative normal acceleration (or velocity in collisions) of the contact coordinate, in which case the links would interpenetrate. Although it is difficult to prove that this situation never occurs in general cases, we have never experienced unrealistic behaviors so far in a number of simulations. Even if interpenetrations occur, they can be recovered by giving a nonzero value to $\ddot{\boldsymbol{r}}_{Rv}$ in Eq.(5.14).
- Non-flat Surfaces
  We cannot completely handle non-flat surfaces where the normal vectors are not uniform in the contact area because the validity check is applied to the sum of the contact forces at all contact points. Summing up the contact forces into a single set of forces and moments eliminates the direction information of the original contact forces, which is essential for the validity check because the conditions are completely different for each direction. A solution for this problem would be to set multiple contact points for one contact pair, in which case we encounter the problem of indeterminate contact forces.

5.6 Collision

The method for simulating collisions is derived by applying the same extension to the method for computing the discontinuous changes of the joint velocities using Newton's Impact Law and conservation of momentum (see Chapter 3).

The conservation of momentum is described as

$$\boldsymbol{H}_C^T \boldsymbol{F}_C = -\boldsymbol{A}\Delta\dot{\boldsymbol{\theta}}_J \tag{5.31}$$

where $\boldsymbol{F}_C$ is the impulse and $\Delta\dot{\boldsymbol{\theta}}_J$ is the change of the joint velocities. The kinematic constraint after the collision is described as

$$\boldsymbol{H}_C(\dot{\boldsymbol{\theta}}_J + \Delta\dot{\boldsymbol{\theta}}_J) = \boldsymbol{v}_C \tag{5.32}$$

where $\boldsymbol{v}_C$ is the velocity of the contact coordinate after the collision.

$\boldsymbol{v}_C$ is computed using Newton's Impact Law. Newton's Impact Law describes the relationship of the normal relative velocities before and after the collision. In order to handle the impulses in tangential and rotational directions, we extend the discussion to all constrained directions. The modified impact law is written as

$$\boldsymbol{v}_C = -\boldsymbol{e}\boldsymbol{v}_{C0} \tag{5.33}$$

where $\boldsymbol{v}_{C0}$ is the vector of relative velocities of constrained directions before the collision and $\boldsymbol{e}$ is a diagonal matrix containing the coefficients of restitution in the constrained directions.

Based on the observation that most contacts in human motions are inelastic, the element of $\boldsymbol{e}$ corresponding to the normal direction is set to 0, namely the links stay in contact after the collision, although it may be set to nonzero values in other situations. If the tangential directions v_x and v_y and the rotational directions ω_x, ω_y and ω_z are constrained, we set the corresponding elements of $\boldsymbol{e}$ to 0, which means that the links stop slipping or rotating if the tangential or rotational impulses were large enough. Therefore, in contrast to contact, even if the links had tangential or angular velocities before a collision, we start by setting all constraints possible for the contact state shown in Table 5.1.

The validity check is applied to $\boldsymbol{F}_C$ using the same method as described in Section 5.5.3, although this is not an exact model. For example, a positive normal impulse only guarantees that the integrated normal force during the collision phase is positive. We might have negative contact force during the collision, in which case the links would separate and the actual normal impact force becomes smaller than the computed one. However, we ignore this special case and apply the same validity check as in contact.

5.7 Simulation Examples

The proposed algorithm was implemented as a private library as well as the computational engine of the humanoid simulator OpenHRP [20, 30, 66]. Information on OpenHRP is available at
http://www.is.aist.go.jp/humanoid/openhrp/English/indexE.html.

5.7.1 Blocks

Fig. 5.3 shows snapshots from the dynamics simulation of colliding blocks on OpenHRP. The thin lines represent the normal vectors at each contact points. RAPID [29] is used for collision detection.

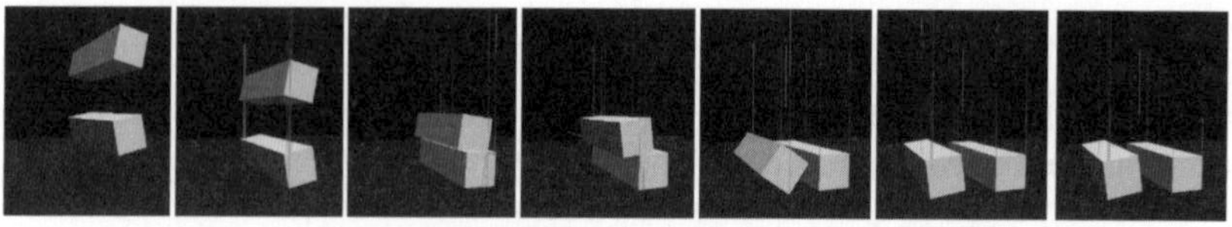

Fig. 5.3. Simulation of colliding blocks.

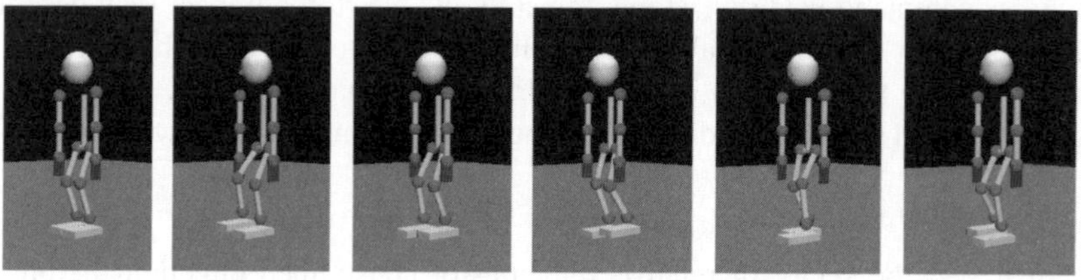

Fig. 5.4. Simulation of a walking humanoid: success.

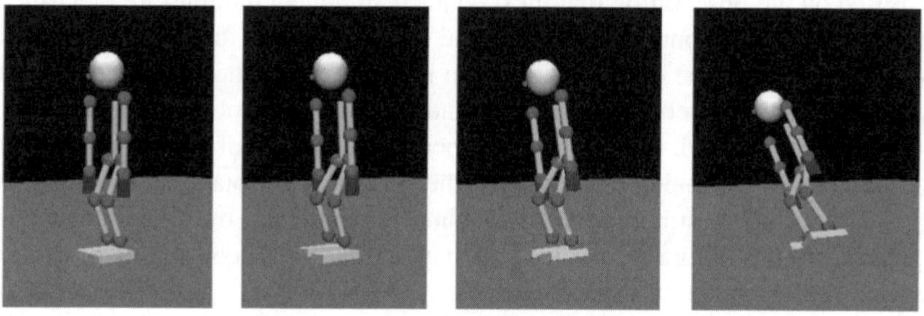

Fig. 5.5. Simulation of a walking humanoid: failure.

5.7.2 Walk

Fig. 5.4 shows the snapshots from a simulation of a walking humanoid robot. The joints are controlled by ideal actuators that exactly follow the desired trajectory. The walking pattern is designed such that the ZMP always lies inside the footprint. In this simulation setup, therefore, the robot is expected to walk successfully without any online controller. However, if we change the mass of the right hand, the robot would fall down as shown in Fig. 5.5.

5.8 Summary

This chapter presented an analytical collision/contact model where the contact forces and constraint conditions are computed by an iterative procedure. The procedure repeats setting a constraint condition and computing the contact forces to finally find the set of constraint condition and contact forces that satisfy the unilateral constraints. The contributions of this chapter are summarized as follows:

(1) The dynamics computation method in Chapter 2 was extended to handle unilateral constraints by enabling explicit computation of constraint forces. This is enabled by modeling the joint between the two links in contact as a 6-DOF free joint. This scheme also eliminates the computation the degrees of freedom and reselecting the generalized coordinates even if the constraint condition changes.
(2) An iterative trial-and-error procedure was proposed to find a set of contact forces and constraint conditions that satisfy the unilateral conditions. Since the computation required for trying a new constraint condition is quite small, the overall method achieves higher efficiency than applying optimization process.
(3) The collision/contact model presented here includes the Coulomb's friction model with both static and dynamic frictions. Static friction is modeled by constraining the tangential motion of the links. The effect of the dynamic friction should be included in the equation of motion by modifying the Jacobian matrix because the slip directions are not constrained.

6

Implicit Integration for Soft Collision/Contact Model

6.1 Introduction

The previous chapter described a collision/contact model for rigid links without deformation. It is reported that, however, the elasticity of the sole plays an important role in maintaining the stability of biped locomotion. It is therefore essential for the dynamics simulation system to include soft contact model.

The advantage of soft contact models is that the constraint forces are computed directly from the current state. The constraint forces can be regarded as external forces applied to the links in contact and we do not have to handle closed kinematic chains. Therefore, the computation time for each step is generally much smaller than rigid contact models. The problem of soft contact models is, however, that it tends to involve impulsive contact forces that requires extremely small integration time step in order to obtain realistic results.

This chapter presents a soft contact model that enables relatively large time step by applying *implicit integration* technique. The parallel forward dynamics algorithm presented in Chapter 4 is applied to the rigid joint constraints of the human figure. The implicit integration technique enables us to use relatively large time step—typically 1 to 4 ms depending on the average velocity of the motion—in spite of the large coefficients of contact springs and dampers. The simulation system can simulate the motion of a 40 DOF human figure controlled by balance controller only a few times slower than real time.

This chapter is organized as follows. First, the basic equations of implicit integration and its application to dynamics simulation are briefly introduced in Section 6.2. Then Section 6.3 presents the method to compute the contact force and its partial derivatives for implicit integration. Finally several simulation examples are shown in Section 6.4 as well as numerical data of the contact force during the simulation.

6.2 Implicit Integration

6.2.1 Overview

Implicit integration is a technique commonly used for simulation of stiff systems. It is capable of simulating the motion of systems with many masses connected by stiff springs and dampers without reducing the sampling time. The idea is to consider the spring and damper forces in the next step, rather than just looking at the instantaneous force. Baraff and Witkin [11] applied this technique to dynamics simulation of cloth which is modeled as particles connected by stiff springs and dampers. Although their work also handles contacts between cloth and rigid bodies, implicit integration is not used in the contact model. This section briefly summarizes the equations for implicit integration technique, which are to be used to derive the proposed collision/contact model.

Equation of motion of a multibody system is written as a second-order differential equation

$$\ddot{\boldsymbol{x}} = \boldsymbol{f}(\boldsymbol{x}, \dot{\boldsymbol{x}}) \tag{6.1}$$

where $\boldsymbol{x}$ is the generalized coordinates of the system. In order to apply implicit integration to Eq.(6.1), we first convert it to a first-order differential equation

$$\frac{d}{dt}\begin{pmatrix} \boldsymbol{x} \\ \boldsymbol{v} \end{pmatrix} = \begin{pmatrix} \boldsymbol{v} \\ \boldsymbol{f}(\boldsymbol{x}, \boldsymbol{v}) \end{pmatrix} \tag{6.2}$$

where $\boldsymbol{v} = \dot{\boldsymbol{x}}$.

In computer simulation, we use the discrete-time version of the equation. If we apply simple *explicit* Euler integration , the discrete equation would be

$$\begin{pmatrix} \Delta\boldsymbol{x} \\ \Delta\boldsymbol{v} \end{pmatrix} = h\begin{pmatrix} \boldsymbol{v}_0 \\ \boldsymbol{f}_0 \end{pmatrix} \tag{6.3}$$

where $\boldsymbol{x}_0 = \boldsymbol{x}(t_0), \boldsymbol{v}_0 = \boldsymbol{v}(t_0), \boldsymbol{f}_0 = \boldsymbol{f}(\boldsymbol{x}_0, \boldsymbol{v}_0)$ and h, t_0 are the sampling time and current time, respectively.

The equation for implicit integration is derived from *backward* Euler integration and written in the following form:

$$\begin{pmatrix} \Delta\boldsymbol{x} \\ \Delta\boldsymbol{v} \end{pmatrix} = h\begin{pmatrix} \boldsymbol{v}_0 + \Delta\boldsymbol{v} \\ \boldsymbol{f}(\boldsymbol{x}_0 + \Delta\boldsymbol{x}, \boldsymbol{v}_0 + \Delta\boldsymbol{v}) \end{pmatrix} \tag{6.4}$$

where velocity and acceleration are computed based on the future state. In contrast to the forward version, solving this equation is not straightforward because both sides include the unknowns $\Delta\boldsymbol{x}$ and $\Delta\boldsymbol{v}$. Instead of solving the equation directly, we apply Taylor series expansion of $\boldsymbol{f}$ to obtain the first-order approximation:

$$\boldsymbol{f}(\boldsymbol{x}_0 + \Delta\boldsymbol{x}, \boldsymbol{v}_0 + \Delta\boldsymbol{v}) = \boldsymbol{f}_0 + \frac{\partial \boldsymbol{f}}{\partial \boldsymbol{x}}\Delta\boldsymbol{x} + \frac{\partial \boldsymbol{f}}{\partial \boldsymbol{v}}\Delta\boldsymbol{v}. \tag{6.5}$$

Substituting Eq.(6.5) into Eq.(6.4) and taking the bottom row, we obtain

$$\left(\boldsymbol{I} - h\frac{\partial \boldsymbol{f}}{\partial \boldsymbol{v}} - h^2\frac{\partial \boldsymbol{f}}{\partial \boldsymbol{x}}\right) \Delta \boldsymbol{v} = h\left(\boldsymbol{f}_0 + h\frac{\partial \boldsymbol{f}}{\partial \boldsymbol{x}}\boldsymbol{v}_0\right) \tag{6.6}$$

which gives the modified acceleration (change of velocity) $\Delta \boldsymbol{v}$. This is the basic equation for implicit integration.

To summarize, implicit integration requires the following four steps to for each time step:

(1) evaluate $\boldsymbol{f}_0$
(2) evaluate $\partial \boldsymbol{f}/\partial \boldsymbol{x}$ and $\partial \boldsymbol{f}/\partial \boldsymbol{v}$
(3) compute elements of Eq.(6.6), and
(4) solve Eq.(6.6).

while explicit integration only uses step 1.

6.2.2 Application to Dynamics Simulation

The main source of computational cost for implicit integration is step 4, where we have to solve a linear equation of N unknowns for an N-DOF chain, which requires $O(N^2)$ computations. Even if the simulator utilizes efficient linear-time forward dynamics algorithms [22, 12], the total complexity will be as large as $O(N^2)$ because of the implicit integration. It is therefore not desirable to select the joint angles as the generalized coordinates $\boldsymbol{x}$ for implicit integration.

In order to reduce the computational cost for implicit integration, we utilize the operational space inertia matrix (OSIM) [48] of the contact links. No additional computation is required for OSIM if the forward dynamics algorithms used in the simulator computes OSIM during the dynamics computation [23]. We may also compute OSIM separately using efficient algorithms [16].

Suppose two links p_i and c_i of two independent kinematic chains A and B, respectively, are in contact as shown in Fig. 6.1, and define the contact frame i between the two links. The lower part of Fig. 6.1 divides frame i to clarify the meaning of some variables. The contact can be interpreted as a 6-degrees-of-freedom joint, whose joint velocity $\dot{\boldsymbol{q}}_i \in \boldsymbol{R}^6$ and acceleration $\ddot{\boldsymbol{q}}_i \in \boldsymbol{R}^6$ are defined as the relative velocity and acceleration of link c_i with respect to link p_i at joint i, namely,

$$\dot{\boldsymbol{q}}_i = \dot{\boldsymbol{r}}_{i,c_i} - \dot{\boldsymbol{r}}_{i,p_i} \tag{6.7}$$

$$\ddot{\boldsymbol{q}}_i = \ddot{\boldsymbol{r}}_{i,c_i} - \ddot{\boldsymbol{r}}_{i,p_i} \tag{6.8}$$

where $\dot{\boldsymbol{r}}_{i,p_i} \in \boldsymbol{R}^6$ and $\dot{\boldsymbol{r}}_{i,c_i} \in \boldsymbol{R}^6$ denote the linear and angular velocities of frame i computed from the velocities of link p_i and link c_i, respectively. The joint force/torque of joint i, $\boldsymbol{\tau}_i \in \boldsymbol{R}^6$, is defined as the contact force/torque applied to link c_i.

Let $\boldsymbol{\Phi}_{Ai}$ and $\boldsymbol{\Phi}_{Bi}$ denote the OSIM of joint i with respect to chains A and B, respectively. OSIM relates the external force applied to a link and its acceleration change. Therefore, the accelerations of frame i on links p_i and c_i are written as

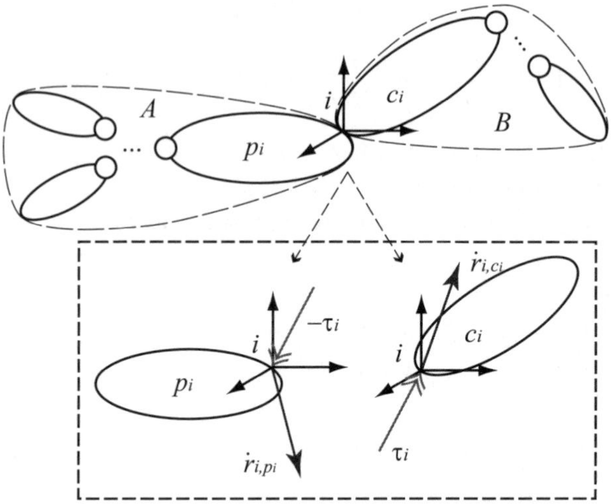

Fig. 6.1. The contact coordinate attached to the two links in contact.

$$\ddot{\boldsymbol{r}}_{i,p_i} = -\boldsymbol{\Phi}_{Ai}\boldsymbol{\tau}_i + \hat{\ddot{\boldsymbol{r}}}_{i,p_i} \tag{6.9}$$

$$\ddot{\boldsymbol{r}}_{i,c_i} = -\boldsymbol{\Phi}_{Bi}\boldsymbol{\tau}_i + \hat{\ddot{\boldsymbol{r}}}_{i,c_i} \tag{6.10}$$

where $\hat{\ddot{\boldsymbol{r}}}_{i,p_i}$ and $\hat{\ddot{\boldsymbol{r}}}_{i,c_i}$ are the accelerations of links p_i and c_i without contact. Note that link p_i receives $-\boldsymbol{\tau}_i$ as the reaction of the contact force/torque. The joint acceleration of joint i is computed by

$$\begin{aligned}\ddot{\boldsymbol{q}}_i &= \ddot{\boldsymbol{r}}_{i,c_i} - \ddot{\boldsymbol{r}}_{i,p_i} \\ &= (\boldsymbol{\Phi}_{Ai} + \boldsymbol{\Phi}_{Bi})\boldsymbol{\tau}_i - \hat{\ddot{\boldsymbol{r}}}_{i,p_i} + \hat{\ddot{\boldsymbol{r}}}_{i,c_i}\end{aligned} \tag{6.11}$$

where Eqs.(6.9) and (6.10) were used. Eq.(6.11) corresponds to the equation of motion Eq.(6.1) used for implicit integration. Therefore, the equation of motion $\boldsymbol{f}(\boldsymbol{q}_i, \dot{\boldsymbol{q}}_i)$ for implicit integration can be written as

$$\boldsymbol{f}(\boldsymbol{q}, \dot{\boldsymbol{q}}) = (\boldsymbol{\Phi}_{Ai} + \boldsymbol{\Phi}_{Bi})\boldsymbol{\tau}_i - \hat{\ddot{\boldsymbol{r}}}_{i,p_i} + \hat{\ddot{\boldsymbol{r}}}_{i,c_i}. \tag{6.12}$$

Implicit integration requires the derivatives of $\boldsymbol{f}$ with respect to $\boldsymbol{q}_i$ and $\dot{\boldsymbol{q}}_i$. The derivatives of $\boldsymbol{\tau}_i$ would be easily computed if it is formulated as a function of relative position and velocity of the links. However, it is difficult to compute the partial derivatives of $\boldsymbol{\Phi}_{Ai}$, $\boldsymbol{\Phi}_{Bi}$, $\hat{\ddot{\boldsymbol{r}}}_{i,p_i}$, and $\hat{\ddot{\boldsymbol{r}}}_{i,c_i}$. We simply ignore these derivatives because their effect is much smaller than those of the contact force.

We first compute joint (contact) force/torque by the method described in the next section. Then we apply implicit integration technique to Eq.(6.12) and compute the actual joint acceleration $\hat{\ddot{\boldsymbol{q}}}_i$. If the actual contact force/torque $\hat{\boldsymbol{\tau}}_i$ is required by the user or the simulation engine, it can be computed by

$$\hat{\boldsymbol{\tau}}_i = (\boldsymbol{\Phi}_{Ai} + \boldsymbol{\Phi}_{Bi})^{-1}(\hat{\ddot{\boldsymbol{q}}}_i + \hat{\ddot{\boldsymbol{r}}}_{i,p_i} - \hat{\ddot{\boldsymbol{r}}}_{i,c_i}). \tag{6.13}$$

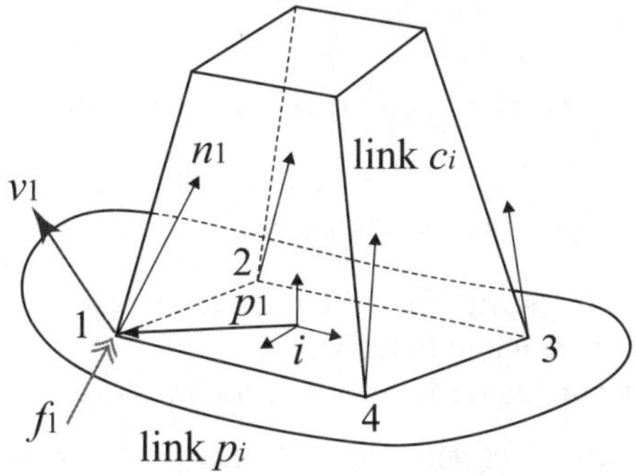

Fig. 6.2. Contact points and associated variables.

6.3 Computation of Contact Force and Its Derivatives

Suppose n_C contact points were detected between links p_i and c_i. For each contact point k $(k = 1, 2, \ldots n_C)$, we define the contact point frame whose z axis is parallel to the normal vector and x, y axes are selected to form a right-handed coordinate. The normal vectors at the contact points may vary for non-flat surfaces. Following variables are defined for contact point k:

$$
\begin{aligned}
{}^i\boldsymbol{x}_k, {}^i\boldsymbol{y}_k, {}^i\boldsymbol{z}_k &: x,\ y,\ z \text{ axes of point } k \text{ frame} \\
{}^i\boldsymbol{p}_k, {}^k\boldsymbol{p}_k &: \text{vector from joint } i \text{ to point } k \\
{}^k\boldsymbol{f}_k = (f_{kx}\ f_{ky}\ f_{kz})^T &: \text{contact force} \\
{}^k\boldsymbol{f}_{Fk} = (f_{kx}\ f_{ky})^T &: \text{friction force} \\
{}^i\boldsymbol{v}_k, {}^k\boldsymbol{v}_k &: \text{relative linear velocity at point } k \\
u_k &: \text{normal velocity} \\
{}^i\boldsymbol{s}_k, {}^k\boldsymbol{s}_k &: \text{slip velocity} \\
d_k &: \text{penetration depth}
\end{aligned}
$$

where the left superscript i or k indicates whether the vector is expressed in joint i or contact point k frame, respectively. We may omit the superscript when the frame does not matter. Fig. 6.2 shows the case of $n_C = 4$.

6.3.1 Normal Force

The normal contact force f_{kz} at contact point k is computed by

$$f_{kz} = (k_P d_k - k_D u_k)/n_C. \tag{6.14}$$

where division by n_C is applied to provide the same vertical force regardless of the density of the polygons.

6.3.2 Friction Force

Following Coulomb's friction law, the friction force would be described as

$$\begin{cases} {}^k\boldsymbol{f}_{Fk} = \hat{\boldsymbol{f}}_{Fk} & \text{if } |\boldsymbol{s}_k| = 0 \text{ and } |\hat{\boldsymbol{f}}_{Fk}| \leq \mu_S f_{kz} \\ {}^k\boldsymbol{f}_{Fk} = -\mu_D f_{kz}{}^k\boldsymbol{s}_k/|\boldsymbol{s}_k| & \text{otherwise} \end{cases} \tag{6.15}$$

where μ_S and μ_D are the static and dynamic friction coefficients respectively, and $\hat{\boldsymbol{f}}_{Fk}$ is the friction force required to realize ${}^k\dot{\boldsymbol{s}}_k = \boldsymbol{O}$.

This model is not suitable for implementation for the following reasons:

- In practice, it is difficult to determine whether the links are slipping or not. The only solution is to set a small threshold that divides static and dynamic friction states.
- Computation of static friction requires either an optimization or a recursive process [9] because the model includes both kinematic and dynamic constraints.
- If only the first condition for static friction was satisfied, the new slip direction is not determined. friction.

Another reason is due to the discrete integration and would be illustrated by the following example. Suppose two links came into contact with large normal relative velocity but small tangential (slip) velocity. Following Coulomb's law, the dynamic friction works between the links to stop slipping and, because of the large normal force and small tangential velocity, the contact will switch to static friction state soon after the collision. If the transition happens before the next time step, however, the links will start slipping in the opposite direction at the next time step, which is not physically realistic.

To avoid these problems, we introduce the following contact model:

(1) Static friction
First compute $\hat{\boldsymbol{f}}_{Fk}$ by

$$\hat{\boldsymbol{f}}_{Fk} = \left(k_{FP}({}^k\hat{\boldsymbol{p}}_k - {}^k\boldsymbol{p}_k) - k_{FD}{}^k\boldsymbol{s}_k\right)/n_C \tag{6.16}$$

where k_{FP} and k_{FD} are positive constant gains and ${}^k\hat{\boldsymbol{p}}_k$ is the position where contact point k is fixed by the static friction. ${}^k\hat{\boldsymbol{p}}_k$ is initialized by the position of point k when the links came into contact, and updated at each time step as described below. This static friction is applied if $||\hat{\boldsymbol{f}}_{Fk}|| \leq \mu_S f_{kz}$. Otherwise, we update ${}^k\hat{\boldsymbol{p}}_k$ by ${}^k\hat{\boldsymbol{p}}_k = {}^k\boldsymbol{p}_k$ and proceed to the next step to compute the dynamic friction.

(2) Dynamic friction
Dynamic friction is computed by

$${}^k\boldsymbol{f}_{Fk} = -\mu_D f_{kz}\frac{w(|\boldsymbol{s}_k|)}{|\boldsymbol{s}_k|}{}^k\boldsymbol{s}_k \tag{6.17}$$

where $w(x)$ $(x \geq 0)$ is a continuous weighting function that satisfies $w(0) = 0$ and $\lim_{x\to+\infty} w(x) = 1$. Note that Eq.(6.17) is equivalent to Coulomb's dynamic friction except for the weighting function.

Each contact point always tries to return to the original position by Eq.(6.16) regardless of its velocity, as long as the static friction does not exceed its limit $\mu_S f_{kz}$. Once the static friction limitation is violated, the contact switches to dynamics friction and the original position is updated at each time step until $\hat{\boldsymbol{f}}_{Fk}$ becomes small enough.

The weighting function w is the key of this friction model. In our experiments we use

$$w(x) = 1 - e^{-kx} \tag{6.18}$$

where k is a positive constant, although other functions may work as well. The problem of switching between static and dynamic friction comes from the fact that the unit vector along the slip direction is not defined when $|\boldsymbol{s}_k| = 0$. With the weighting function, in contrast, the dynamic friction is continuous even near $|\boldsymbol{s}_k| = 0$ because

$$\lim_{x \to 0} \frac{w(x)}{x} = k. \tag{6.19}$$

We can therefore approximate the dynamic friction at small $|\boldsymbol{s}_k|$ by

$${}^k\boldsymbol{f}_{Fk} = -\mu_D f_{kz} k \, {}^k\boldsymbol{s}_k. \tag{6.20}$$

The weighting function also solves the problem from discrete integration because it effectively reduces the friction force when the slip velocity is not large enough and prevent unrealistic behavior.

6.3.3 Transformation to Joint Force/Torque

The point contact forces are transformed to joint force/torque by

$$\boldsymbol{\tau}_i = \sum_k \boldsymbol{J}_k^T \, {}^k\boldsymbol{f}_k. \tag{6.21}$$

where $\boldsymbol{J}_k \in \boldsymbol{R}^{3\times 6}$ is the Jacobian matrix defined as

$$\boldsymbol{J}_k \stackrel{\triangle}{=} \frac{\partial \, {}^k\boldsymbol{p}_k}{\partial \boldsymbol{q}_i}. \tag{6.22}$$

Let ${}^k\boldsymbol{v}_{k,m}$ $(m = p_i, c_i)$ denote the linear velocity of point k when it is assumed to be fixed to link m. $\boldsymbol{J}_k$ is the Jacobian matrix that relates ${}^k\boldsymbol{v}_{k,m}$ and ${}^i\dot{\boldsymbol{r}}_{i,m}$ as

$${}^k\boldsymbol{v}_{k,m} = \boldsymbol{J}_k {}^i\dot{\boldsymbol{r}}_{i,m}. \tag{6.23}$$

We can use the same Jacobian matrix for both links because they instantaneously share the joint i and point k frames. Subtracting Eq.(6.23) for $m = p_i$ from that for $m = c_i$, we obtain

$${}^k\boldsymbol{v}_{k,c_i} = \boldsymbol{J}_k({}^i\dot{\boldsymbol{r}}_{i,c_i} - {}^i\dot{\boldsymbol{r}}_{i,p_i}) \tag{6.24}$$

which is then rewritten as

$$^{k}\boldsymbol{v}_k = \boldsymbol{J}_k \dot{\boldsymbol{q}}_i \tag{6.25}$$

using the definitions of $^{k}\boldsymbol{v}_k$ and $\dot{\boldsymbol{q}}_i$.

In order to compute $\boldsymbol{J}_k$, let us first divide it as $\boldsymbol{J}_k = {}^{i}\boldsymbol{R}_k^{T}\,{}^{i}\boldsymbol{J}_k$ where $^{i}\boldsymbol{J}_k$ satisfies

$$^{i}\boldsymbol{v}_k = {}^{i}\boldsymbol{J}_k \dot{\boldsymbol{q}}_i. \tag{6.26}$$

$^{i}\boldsymbol{J}_k$ can be intuitively formed as

$$^{i}\boldsymbol{J}_k = \left(\boldsymbol{I}\ [-{}^{i}\boldsymbol{p}_k \times]\right) \tag{6.27}$$

where $\boldsymbol{I} \in \boldsymbol{R}^{3\times 3}$ is the identity matrix and $[\boldsymbol{x}\times]$ denotes the cross product matrix of $\boldsymbol{x}$. Therefore, $\boldsymbol{J}_k$ is computed by

$$\begin{aligned} \boldsymbol{J}_{k,m} &= \left({}^{i}\boldsymbol{R}_k^{T}\ {}^{i}\boldsymbol{R}_k^{T}[-{}^{i}\boldsymbol{p}_k \times]\right) \\ &= \begin{pmatrix} {}^{i}\boldsymbol{x}_k^{T} & ({}^{i}\boldsymbol{p}_k \times {}^{i}\boldsymbol{x}_k)^{T} \\ {}^{i}\boldsymbol{y}_k^{T} & ({}^{i}\boldsymbol{p}_k \times {}^{i}\boldsymbol{y}_k)^{T} \\ {}^{i}\boldsymbol{z}_k^{T} & ({}^{i}\boldsymbol{p}_k \times {}^{i}\boldsymbol{z}_k)^{T} \end{pmatrix}. \end{aligned} \tag{6.28}$$

6.3.4 Derivatives

Next we compute the partial derivatives of $\boldsymbol{\tau}_i$. First let us consider $\partial \boldsymbol{\tau}_i / \partial \boldsymbol{q}_i$. From Eqs.(6.22) and (6.21) we know that

$$\begin{aligned} \frac{\partial \boldsymbol{\tau}_i}{\partial \boldsymbol{q}_i} &= \sum \boldsymbol{J}_k^{T} \frac{\partial\, {}^{k}\boldsymbol{f}_k}{\partial \boldsymbol{q}_i} \\ &= \sum \boldsymbol{J}_k^{T} \frac{\partial\, {}^{k}\boldsymbol{f}_k}{\partial\, {}^{k}\boldsymbol{p}_k} \frac{\partial\, {}^{k}\boldsymbol{p}_k}{\partial \boldsymbol{q}_i} \\ &= \sum \boldsymbol{J}_k^{T} \boldsymbol{F}_{kp} \boldsymbol{J}_k \end{aligned} \tag{6.29}$$

where $\boldsymbol{F}_{kp} \triangleq \partial\, {}^{k}\boldsymbol{f}_k / \partial\, {}^{k}\boldsymbol{p}_k$ and the partial derivative of $\boldsymbol{J}_k$ is ignored. $\boldsymbol{F}_{kp}$ indicates how the contact force at point k changes as the contact point moves.

If static friction is effective, $\boldsymbol{F}_{kp}$ is computed as

$$\boldsymbol{F}_{kp} = \begin{pmatrix} -k_{FP} & 0 & 0 \\ 0 & -k_{FP} & 0 \\ 0 & 0 & -k_P \end{pmatrix} \tag{6.30}$$

while if dynamic friction is effective $\boldsymbol{F}_{kp}$ becomes

$$\boldsymbol{F}_{kp} = \begin{pmatrix} 0\ 0 & \frac{\partial\, {}^{k}\boldsymbol{f}_{Fk}}{\partial d_k} \\ 0\ 0 & \\ 0\ 0 & -k_P \end{pmatrix} \tag{6.31}$$

$$\frac{\partial\, {}^{k}\boldsymbol{f}_{Fk}}{\partial d_k} = \mu_D k_P \frac{w(|\boldsymbol{s}_k|)}{|\boldsymbol{s}_k|}\, {}^{k}\boldsymbol{s}_k. \tag{6.32}$$

If $|\boldsymbol{s}_k|$ is small, Eq.(6.32) is replaced by

$$\frac{\partial\, {}^k\boldsymbol{f}_{Fk}}{\partial d_k} = \mu_D k_P k\, {}^k\boldsymbol{s}_k. \tag{6.33}$$

Similarly, $\partial\boldsymbol{\tau}_i/\partial\dot{\boldsymbol{q}}_i$ is computed by

$$\begin{aligned}\frac{\partial\boldsymbol{\tau}_i}{\partial\dot{\boldsymbol{q}}_i} &= \sum \boldsymbol{J}^T_{k,p_i}\frac{\partial\, {}^k\boldsymbol{f}_k}{\partial\dot{\boldsymbol{q}}_i} \\ &= \sum \boldsymbol{J}^T_{k,p_i}\frac{\partial\, {}^k\boldsymbol{f}_k}{\partial\, {}^k\boldsymbol{v}_k}\frac{\partial\, {}^k\boldsymbol{v}_k}{\partial\dot{\boldsymbol{q}}_i} \\ &= \sum \boldsymbol{J}^T_{k,p_i}\boldsymbol{F}_{kd}\boldsymbol{J}_{k,p_i}\end{aligned} \tag{6.34}$$

where $\boldsymbol{F}_{kd} \stackrel{\triangle}{=} \partial\, {}^k\boldsymbol{f}_k/\partial\, {}^k\boldsymbol{v}_k$. If static friction is effective, $\boldsymbol{F}_{kd}$ is

$$\boldsymbol{F}_{kd} = \begin{pmatrix} -k_{FD} & 0 & 0 \\ 0 & -k_{FD} & 0 \\ 0 & 0 & -k_D \end{pmatrix} \tag{6.35}$$

while if dynamic friction is effective $\boldsymbol{F}_{kd}$ becomes

$$\boldsymbol{F}_{kd} = \begin{pmatrix} \frac{\partial\, {}^k\boldsymbol{f}_{Fk}}{\partial\, {}^k\boldsymbol{s}_k} & \frac{\partial\, {}^k\boldsymbol{f}_{Fk}}{\partial u_k} \\ 0\;0 & -k_D \end{pmatrix} \tag{6.36}$$

$$\frac{\partial\, {}^k\boldsymbol{f}_{Fk}}{\partial\, {}^k\boldsymbol{s}_k} = -\mu_D f_{kz}\left\{\frac{w(|\boldsymbol{s}_k|)}{|\boldsymbol{s}_k|}\boldsymbol{I} + \left(ke^{-k|\boldsymbol{s}_k|} - \frac{w(|\boldsymbol{s}_k|)}{|\boldsymbol{s}_k|}\right)\frac{{}^k\boldsymbol{s}_k\,{}^k\boldsymbol{s}_k^T}{|\boldsymbol{s}_k|^2}\right\} \tag{6.37}$$

$$\frac{\partial\, {}^k\boldsymbol{f}_{Fk}}{\partial u_k} = \mu_D k_D\frac{w(|\boldsymbol{s}_k|)}{|\boldsymbol{s}_k|}\, {}^k\boldsymbol{s}_k \tag{6.38}$$

where $\boldsymbol{I} \in \boldsymbol{R}^{2\times 2}$ is the identity matrix. We used the following relationships to derive Eqs.(6.36)–(6.38):

$$\frac{\partial w}{\partial|\boldsymbol{s}_k|} = ke^{-k|\boldsymbol{s}_k|} \tag{6.39}$$

$$\frac{\partial|\boldsymbol{s}_k|}{\partial\boldsymbol{s}_k} = \frac{\partial\boldsymbol{s}_k^T}{\partial|\boldsymbol{s}_k|}. \tag{6.40}$$

If $|\boldsymbol{s}_k|$ is small, we use

$$\frac{\partial\, {}^k\boldsymbol{f}_{Fk}}{\partial^k\boldsymbol{s}_k} = -\mu_D f_{kz} k\boldsymbol{I} \tag{6.41}$$

$$\frac{\partial\, {}^k\boldsymbol{f}_{Fk}}{\partial u_k} = \mu_D k_D k\, {}^k\boldsymbol{s}_k. \tag{6.42}$$

6.3.5 Summary of the Method

The proposed method is summarized by the following four steps:

(1) Compute the contact force, its derivatives, and the Jacobian matrix at each contact point.
(2) Transform the contact force and the derivatives to the joint coordinate.
(3) Compute the joint acceleration using implicit integration technique.
(4) Compute the new contact force.

6.3.6 Determining the Parameters

Our contact model uses the following constant parameters:

(1) Position and velocity gains for the normal force, k_P and k_D.
(2) Static and dynamic friction coefficients, μ_S and μ_D.
(3) Constant for the weighting function, k.
(4) Position and velocity gains for the friction force, k_{FP} and k_{FD}.

Parameters 1. and 2. are determined from the physical properties of the materials and can be measured through experiments. The other three parameters are introduced only to solve the problem of discrete-time integration and Coulomb's friction model and, therefore, should not affect the simulation results.

The constant parameter for the weighting function determines the difference between the weighted friction force and the friction force of standard Coulomb's friction law. The role of the weighting function is to realize continuous transition between static and dynamic friction states under discrete-time integration. Therefore, smaller integration time step will allow larger k and the model will be closer to Coulomb's model. The most simple function to formulate this relationship would be $k = c/\Delta t$ where c is a constant and Δt is the integration time step.

The position and velocity gains for the friction force k_{FP} and k_{FD} determine the maximum slip velocity at which the static friction can be active. Note that, unlike explicitly specifying a constant slip velocity threshold to apply static friction, the maximum velocity also depends on the normal force f_{kz}. One practical way to determine these parameters would be: (1) first determine k_{FP} so that static friction is applied at model-dependent average vertical force and user-defined maximum position error, and (2) determine k_{FD} through simple simulations like those in Section 6.4.1 such that the motion is not over- or under-damped. As shown in the examples section, the results match the analytical solutions for various environments with two different sets of k_{FP} and k_{FD}. The experiments also prove that these parameters do not greatly affect the simulation results.

6.4 Simulation Examples

The method is implemented on top of our the forward dynamics algorithm presented in Chapter 4 and a freely available collision detection library ColDet[1] to which the

[1] http://photoneffect.com/coldet/

author added a simple code to compute the normal vector and penetration depth. The enhanced collision detection library analyzes VRML format files to obtain the three-dimensional shape of the links and is capable of handling generic polygonal shapes. Our code works well for all of the cases we have tried with relatively simple geometry, although integrating with sophisticated algorithms for computing the penetration depth (e.g. [41]) will be included in the future work. Our contact model may also work with other dynamics simulation engines as long as they provide OSIM or at least the information required to compute OSIM.

In the following examples, we used $k_P = 1 \times 10^5$ and $k_D = 1 \times 10^3$ except where otherwise noted. We determined the value of k by $k = 0.1/\Delta t$ where Δt is the integration time step.

6.4.1 Comparison with Analytical Solution

In order to compare our friction model with widely accepted Coulomb's model, we simulated the motion of a box slipping on a floor with different initial velocities, for which we can easily compute the analytical solutions using Coulomb's friction model.

Fig. 6.3 shows the trajectories of a box when it started slipping with initial velocities 1 m/s and 2 m/s on a flat floor. We used two different sets of friction coefficients $(\mu_S, \mu_D) = (0.35, 0.25), (0.65, 0.55)$ and two sets of position and velocity gains for friction force $(k_{FP}, k_{FD}) = (1000, 700), (2000, 1400)$. The results prove that the proposed friction model approximates Coulomb's model to a practically sufficient level, and that values of k_{FP} and k_{FD} do not greatly affect the result.

Fig. 6.4 shows the trajectories on a slope of 30 degrees. Analytically, the box will come to a complete stop if $\mu_D < \sqrt{3}/3 = 0.577$, and continue accelerating otherwise. The friction coefficients used in these examples are $(\mu_S, \mu_D) = (0.64, 0.54), (0.68, 0.58), (0.72, 0.62)$. These results also show good match with the analytical solution.

6.4.2 Comparison with Explicit Integration

Implicit integration was compared with explicit Euler and 4th-order Runge-Kutta integrations using a simple setting where a box falls down onto a flat floor. The results are shown in Table 6.1 along with the computation times per frame measured on a PC with a Pentium III 1GHz processor. Figs. 6.5 (a)–(e) show the snapshots taken every 0.1s in several selected cases. In Table 6.1, the cells with ○ indicates that the motion of the box was correctly simulated as in Fig. 6.5 (a), while × indicates that the box flew away immediately after touching the floor due to the integration error as in Fig. 6.5 (b). In cases marked with △, the box did remain on the floor but the results were obviously incorrect, as shown in Figs. 6.5 (c)–(e).

In this particular case, implicit integration allows approximately 16 times larger time step than explicit Euler integration and 4 times larger than explicit 4th-order Runge-Kutta. The total simulation time of implicit integration, therefore, will be reduced to 1/8 of both explicit Euler and 4th-order Runge-Kutta integrations.

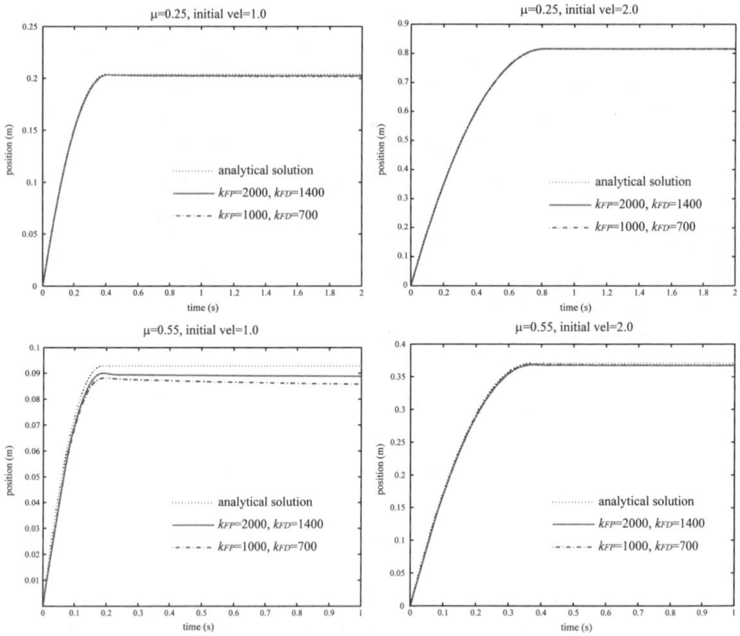

Fig. 6.3. Motion of a box on a horizontal plane.

Table 6.1. Comparison of precision and computation time of implicit and explicit integrations.

time step (ms)	implicit	explicit	
		Euler	Runge-Kutta
8	×(c)	×	×
4	△(b)	×	×
2	○(a)	×	×
1	○	×	△(e)
0.5	○	×	○
0.25	○	△(d)	○
0.125	○	○	○
time	1.05	0.585	2.31

6.4.3 Applications

Figs. 6.6 and 6.7 are the snapshots from simulations of ten dominos with different initial configurations. The integration time step is 2 ms and the computation time was 26.9 ms per frame in both examples on a Pentium IV 2GHz processor. The computation for implicit integration took approximately 10% of the total simualtion time. Most of the time was spent for the forward dynamics computation because they form a complex closed kinematic chain.

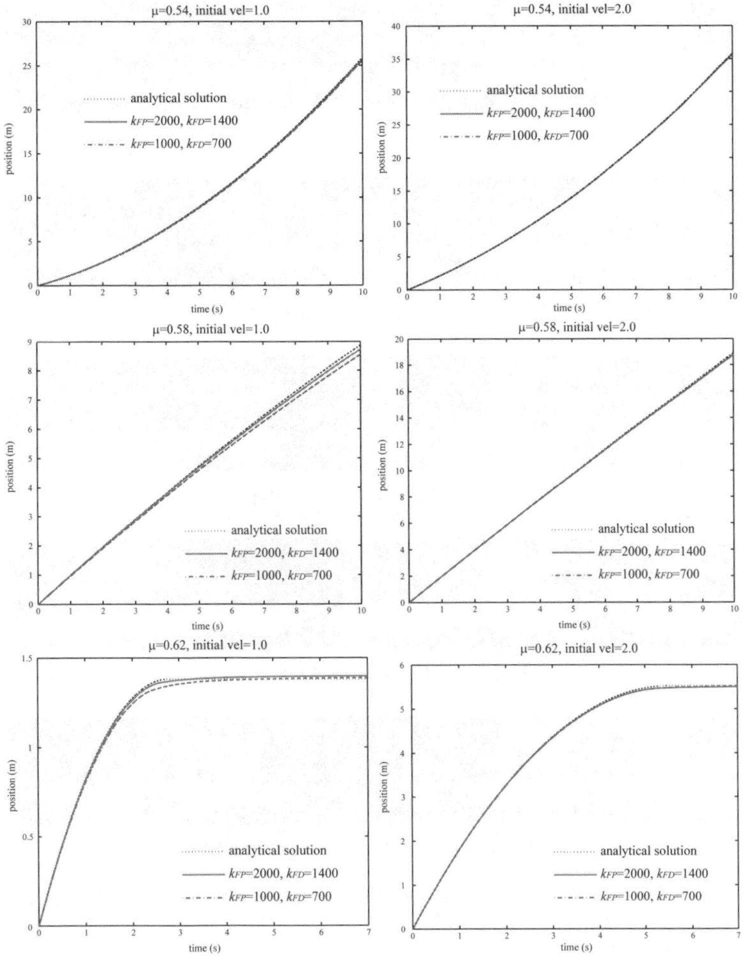

Fig. 6.4. Motion of a box on a slope of 30 degrees.

Fig. 6.8 shows a winding up motion of a 4 m-long serial chain around a cylinder. The parameters used in this example were $(k_P, k_D) = (1 \times 10^6, 1 \times 10^4)$ and $(k_{FP}, k_{FD}) = (2000, 700)$. The collisions between the links of the chain are also considered.

Fig. 6.9 shows the snapshots from a simulation of a walking human figure. The human figure is controlled by two controllers to keep walking. The first controller modifies the pattern to keep the robot balanced [37], while the second controller simply makes the joints follow the reference joint trajectories by PD torque control.

Computation of the joint accelerations takes approximately 4.4 ms for a 40DOF human figure including dynamics simulation, collision detection, control, and graphical representation on a PC with a Pentium IV 2GHz processor. Fig. 6.10 shows

(a) Implicit integration, time step 2 ms

(b) Implicit integration, time step 4 ms

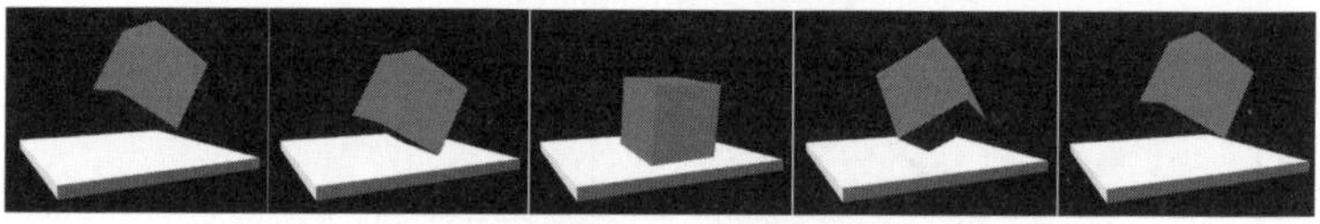

(c) Implicit integration, time step 8 ms

(d) Explicit Euler integration, time step 0.25 ms

(e) Explicit 4th-order Runge-Kutta integration, time step 1 ms

Fig. 6.5. Snapshots from simulations using implicit and explicit integrations.

the normal force computed in the simulation and demonstrates that the change of load distribution between feet is reasonably simulated. The same figure and controller were applied on an floor with obstacles, which ended up in the result shown in Fig. 6.11.

Table 6.2 summarizes the complexity (number of triangles and link pairs for collision detection, total degrees of freedom) and the computation time of each example. The computation times were measured on a Pentium IV 2GHz processor.

6.5 Summary

The contributions of this chapter are summarized as follows:

(1) Implicit integration was applied to a penalty-based contact model in order to enable large time steps. Derivatives of inertial matrices are ignored because their

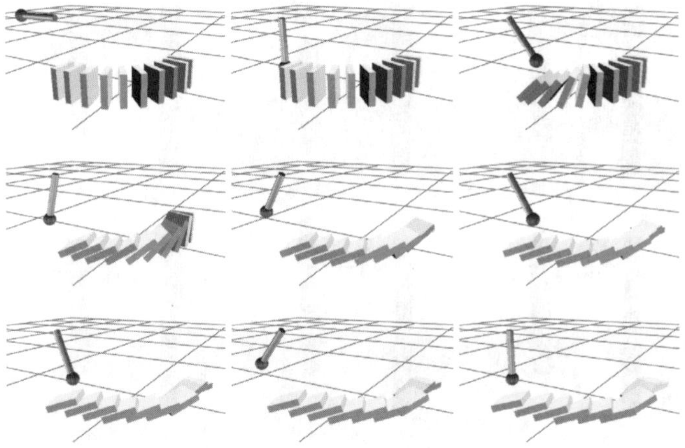

Fig. 6.6. Simulation of domino (1), captured every 0.3 s.

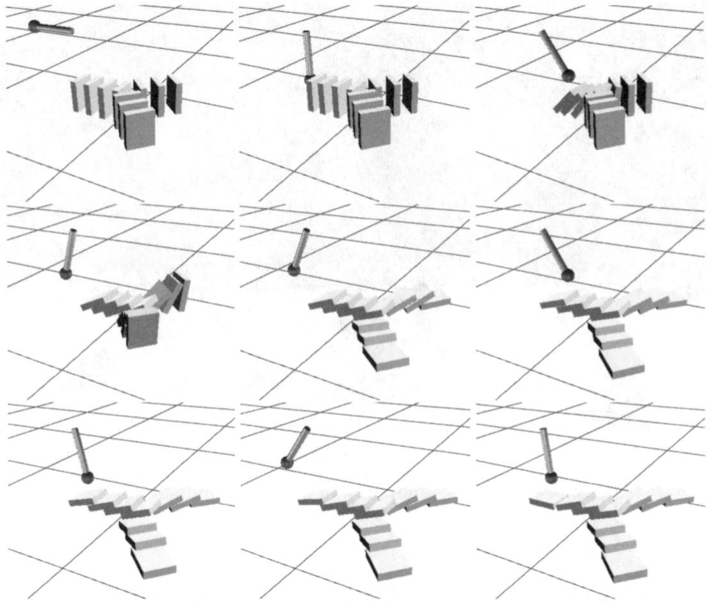

Fig. 6.7. Simulation of domino (2), captured every 0.3 s.

effects are much smaller than those of the contact force generated by stiff springs and dampers.

(2) A new friction model, including static and slip frictions, was proposed. This model solves the numerical problems arising from discrete-time integration and discontinuity of Coulomb's static and dynamic friction models.

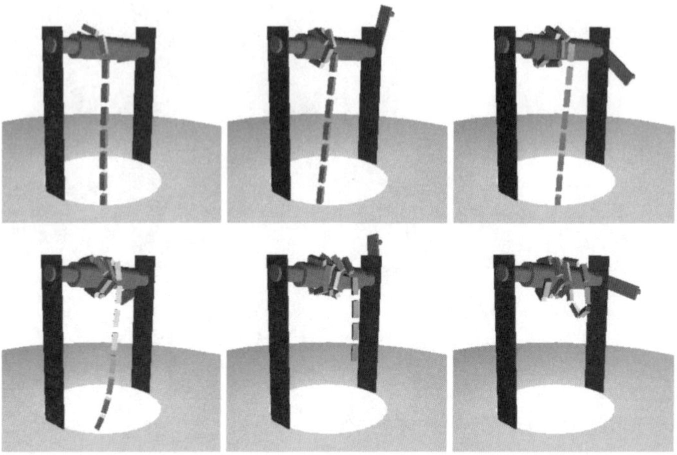

Fig. 6.8. Simulation of winding up a serial chain, captured every 2 s.

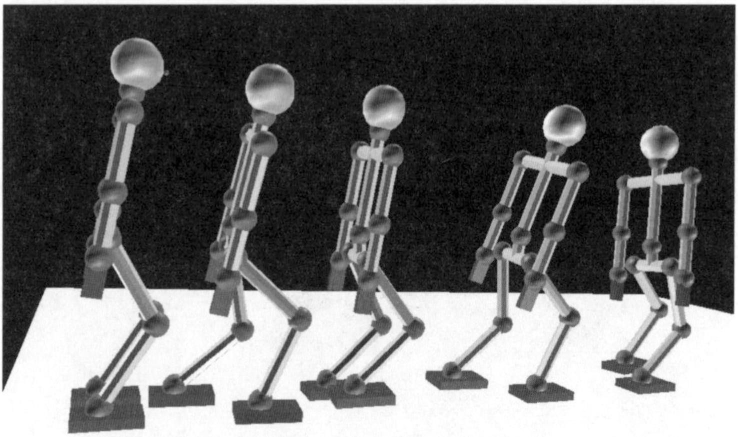

Fig. 6.9. Dynamics simulation of a walking human figure.

(3) Comparison with analytical solutions proved that the proposed friction model effectively approximates Coulomb's friction model.
(4) Comparison with explicit integration proved that the increase of computation time per step due to implicit integration is much smaller than the effect of using larger time steps, that is, the total simulation time is greatly reduced by implicit integration.
(5) Simulation examples using more complex settings demonstrated the power of our model. The computation time for simulation of the motion of human figures is only a few time longer than real time, including computations for simulation,

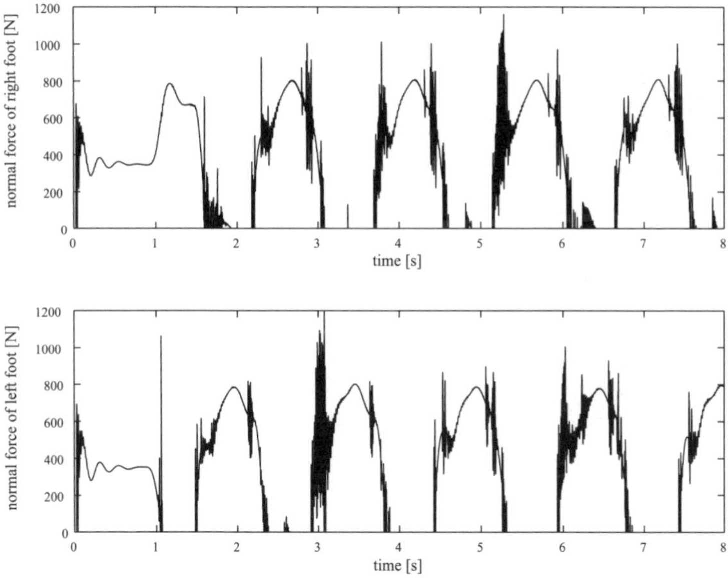

Fig. 6.10. Normal contact force computed in the simulation.

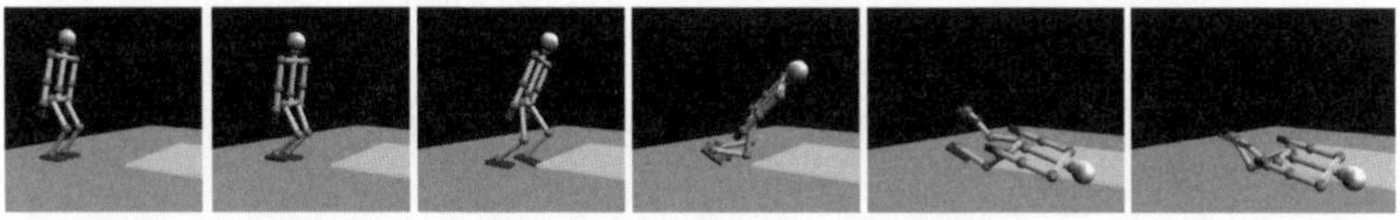

Fig. 6.11. Dynamics simulation of a walking human figure on a floor with obstacles.

Table 6.2. Complexity and computation time of the simulation examples.

example	domino	chain	human
no. of triangles	756	612	2812
no. of pairs	77	211	15
DOF	61	41	40
comp. time (ms)	26.9	2.73	4.36

controller and collision detection. The time step can be as large as 1–4 ms, which is almost ten times larger than typical time steps for penalty-based contact models with explicit integration.

Part II

Motion Generation of Human Figures

7

Interactive Motion Generation via Dynamics Filter

7.1 Introduction

Interactivity is a key issue in many applications of humanoid robots working in a changing or unknown environment with a human. A robot may need to avoid an obstacle suddenly appearing in front of it, resist external forces applied to its body, or abruptly change the walking direction as ordered by its master. Most previous research has discussed the interactivity in humanoid robot motion on the behavioral level. Although such approaches are capable of generating intelligent behaviors, there still remains the problem of attaching physically feasible and natural full-body motion to the behavior.

Physical consistency, the condition that the motion is physically possible for some choice of internal forces, is essential for stable and reliable motions of humanoid robots. The total angular momentum, for example, should be constant while the human figure is in the air. The zero moment point (ZMP) should be in the contact area if one or more links are in contact with the environment. There may be other constraints depending on a specific robot such as joint torque and angle limits.

In this chapter, we propose the concept of a *dynamics filter* and present an implementation with high interactivity. The basic function of the dynamics filter is to convert a physically inconsistent motion into a consistent one. Its main focus is in generating a motion sequence that is theoretically feasible for a given human figure. Any motion data may be input to the filter as a reference: human motion capture data, motions created by hand using animation software, or kinematically synthesized motions. The dynamics filter provides large flexibility in applying the existing motion data to a different model or environment. The implementation of the dynamics filter is based on the method for computing the dynamics of structure-varying kinematic chains presented in Chapter 3 and the rigid contact model in Chapter 5.

This chapter was adapted from, by permission, K. Yamane and Y. Nakamura, "Dynamics Filter—Concept and Implementation of On-Line Motion Generator for Human Figures," IEEE Transactions on Robotics and Automation, vol.19, no.3, pp.421–432, 2003.

This chapter is organized as follows. In Section 7.2 we review the existing techniques for generating motions of human figures. Implementation of the dynamics filter is described in Section 7.3. Some examples of applying the dynamics filter to various situations are also provided in Section 7.3.4. Finally we summarize this chapter in Section 7.4.

7.2 Related Work

Humanoids

Interactivity in the whole-body motion of humanoids is not discussed very much, except for a few works such as [70] where carefully designed motion patterns are mixed according to the user input. However, they are not fully interactive in the sense that they allow only a few limited types of motion.

One of the major approaches is to prepare a physically consistent motion and give it as a reference to the robot controlled by some online controllers. Generation of physically consistent motion is usually done by describing the motion (either as the trajectory of the tip links or the joint angles) by a few parameters and optimizing them using techniques such as ZMP [98, 75], inverted pendulum mode [74, 40], and optimal gradient techniques [61]. In any cases, parameterization scheme depends on the motion, and the optimization is done off-line due to the heavy computational load, allowing robots to make only predefined motions. DasGupta et al. [19], in contrast, used captured human motion data as the initial guess for the optimization based on ZMP constraint. Tak et al. [94] proposed an alternative off-line method for the same purpose in the computer animation context. Online controllers are designed in various ways including heuristic methods [27, 37], learning [36], and limit cycle control [60].

We can find two major problems in current approaches. The first is that they are designed for each specific type of motion. If we would like to generate a different motion, or even just modify the motion slightly, we will have to build a completely different set of pattern generator and controller. The second problem is that parameterization of a motion tends to yield an artificial, unfriendly motion.

Computer Animation

The most popular approach in computer animation, where human-likeliness is the primary requirement, is to use captured human motion data. In fact, many motion libraries and tools for editing and retargetting captured motions have been developed and even commercially available. Much of the recent research focuses on motion editing under the existence of motion clips instead of creating a new motion from scratch. Gleicher [28] used global optimization techniques to retarget a captured motion to human characters with various dimensions. Choi et al. [17], on the other hand, proposed a method for online motion retargetting. Rose and his colleagues [86, 87] studied motion blending problem while preserving the kinematic constraints in the original motions. Lee et al. [47] developed an intuitive interface for editing global trajectory of a captured motion.

Physical consistency is also important for reality of the motion. Ko et al. [43] use inverse dynamics to ensure the physical consistency. Popović et al. [83] proposed a method for realizing physical consistency for different human character using simplified dynamic model. Pollard et al. [81] developed a different approach based on the force information obtained through motion capture.

Another approach is to use integrated control schemes to create various behaviors without captured data. A number of controllers have been developed for various motions such as walking [95], diving [97], bicycling [13], and athletic motions [32]. A problem of this approach is that we need to prepare different controller for each behavior, which leads to difficulty in generating intermediate behaviors. Faloutsos et al. [21] proposed an approach to overcome this problem.

7.3 Dynamics Filter Implementation

7.3.1 Basic Equation

Eq.(5.17) gives the unique solution for the joint accelerations and the constraint forces when the constraint condition and the actuator torques $\boldsymbol{\tau}_A$ are known, which is the case for dynamics simulation. Our purpose in the dynamics filter, on the other hand, is to generate the motion itself based on some optimization scheme while the actuator torques are initially unknown. Moreover, we need an equation with some redundancy for optimization. Taking these facts into account, we modify Eq.(5.17) for the dynamics filter as:

$$\begin{pmatrix} \boldsymbol{A} & -\boldsymbol{H}_C^T & -\boldsymbol{H}_J^T \\ \boldsymbol{H}_C & \boldsymbol{O} & \boldsymbol{O} \end{pmatrix} \begin{pmatrix} \ddot{\boldsymbol{\theta}}_J \\ \boldsymbol{\tau}_C \\ \boldsymbol{\tau}_J \end{pmatrix} = \begin{pmatrix} -\boldsymbol{b} \\ -\dot{\boldsymbol{H}}_C \dot{\boldsymbol{\theta}}_G \end{pmatrix} \tag{7.1}$$

or in a simpler form,

$$\boldsymbol{W}\boldsymbol{x} = \boldsymbol{u} \tag{7.2}$$

where

$$\boldsymbol{W} \triangleq \begin{pmatrix} \boldsymbol{A} & -\boldsymbol{H}_C^T & -\boldsymbol{H}_J^T \\ \boldsymbol{H}_C & \boldsymbol{O} & \boldsymbol{O} \end{pmatrix} \tag{7.3}$$

$$\boldsymbol{x} \triangleq \begin{pmatrix} \ddot{\boldsymbol{\theta}}_J \\ \boldsymbol{\tau}_C \\ \boldsymbol{\tau}_J \end{pmatrix} \tag{7.4}$$

$$\boldsymbol{u} \triangleq \begin{pmatrix} -\boldsymbol{b} \\ -\dot{\boldsymbol{H}}_C \dot{\boldsymbol{\theta}}_G \end{pmatrix}. \tag{7.5}$$

Eq.(7.1) is a redundant linear equation in $(\ddot{\boldsymbol{\theta}}_J \ \boldsymbol{\tau}_C \ \boldsymbol{\tau}_J)^T$, whose solutions represents all the physically feasible motions under the given set of constraints. The role

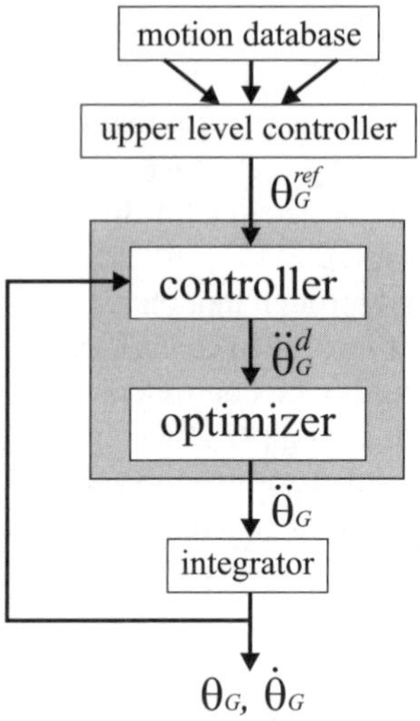

Fig. 7.1. The structure of the dynamics filter.

of the dynamics filter is to select the most appropriate solution of Eq.(7.1) taking various factors into account. As described in Section 5.5, if $\boldsymbol{\tau}_C$ in the solution does not satisfy the unilateral conditions for the contact forces, we change the constraint condition and recompute a new solution.

7.3.2 Outline

Fig. 7.1 shows the structure of the dynamics filter and its typical application, where the shadowed box represents the main part of the dynamics filter. The filter we developed consists of two parts: controller and optimizer. The controller part computes the desired (but not always feasible) joint accelerations considering the reference motion and the current state. This part itself consists of two feedback sections: local and global. The local feedback section simply computes the temporary desired accelerations by the local feedback controller at each joint, which are modified by the global feedback section to maintain the whole-body configuration. Given the desired joint accelerations, the optimization part then computes the solution of Eq.(7.1) that minimizes the error between the actual and desired joint accelerations. Both the controller and optimizer parts are designed to require only the current state and reference motion to enable on-line filtering.

7.3.3 Details

Controller

First, the local feedback section computes the temporary desired acceleration of the generalized coordinates $\ddot{\boldsymbol{\theta}}_G^{d0}$ by simple joint angle and velocity feedback:

$$\ddot{\boldsymbol{\theta}}_G^{d0} = \ddot{\boldsymbol{\theta}}_G^{ref} + \boldsymbol{K}_D(\dot{\boldsymbol{\theta}}_G^{ref} - \dot{\boldsymbol{\theta}}_G) + \boldsymbol{K}_P(\boldsymbol{\theta}_G^{ref} - \boldsymbol{\theta}_G) \tag{7.6}$$

where $\boldsymbol{\theta}_G^{ref}$ is the generalized coordinates in reference motion, and $\boldsymbol{K}_D$ and $\boldsymbol{K}_P$ are gain matrices.

Then, in order to influence the configuration of the whole body, the global feedback section modifies the desired acceleration to feedback the position and orientation of a specified point $\boldsymbol{P}$ in the upper body. The desired acceleration of $\boldsymbol{P}$, $\ddot{\boldsymbol{\theta}}_P^d$, is computed by a similar feedback law as:

$$\ddot{\boldsymbol{r}}_P^d = \ddot{\boldsymbol{r}}_P^{ref} + \boldsymbol{K}_{DP}(\dot{\boldsymbol{r}}_P^{ref} - \dot{\boldsymbol{r}}_P) + \boldsymbol{K}_{PP}(\boldsymbol{r}_P^{ref} - \boldsymbol{r}_P) \tag{7.7}$$

where $\boldsymbol{r}_P^{ref}$ is the position and orientation of $\boldsymbol{P}$ in the reference motion, which can be obtained by the forward kinematics computation, $\boldsymbol{K}_{DP}$ and $\boldsymbol{K}_{PP}$ are gain matrices, and $\boldsymbol{r}_P$ is the current position and orientation of $\boldsymbol{P}$. The temporary desired acceleration of the generalized coordinates $\ddot{\boldsymbol{\theta}}_G^{d0}$ is modified into $\ddot{\boldsymbol{\theta}}_G^d$ so that the desired acceleration of $\boldsymbol{P}$, $\ddot{\boldsymbol{r}}_P^d$, is realized, by

$$\ddot{\boldsymbol{\theta}}_G^d = \ddot{\boldsymbol{\theta}}_G^{d0} + \Delta\ddot{\boldsymbol{\theta}}_G^d \tag{7.8}$$

$$\Delta\ddot{\boldsymbol{\theta}}_G^d = \boldsymbol{J}_P^\sharp(\ddot{\boldsymbol{r}}_P^d - \ddot{\boldsymbol{r}}_P^{d0}) \tag{7.9}$$

where $\ddot{\boldsymbol{r}}_P^{d0} \triangleq \boldsymbol{J}_P\ddot{\boldsymbol{\theta}}_G^{d0} + \dot{\boldsymbol{J}}_P\dot{\boldsymbol{\theta}}_G$, $\boldsymbol{J}_P \triangleq \partial\boldsymbol{J}_P/\partial\boldsymbol{\theta}_G$, and $\boldsymbol{J}_P^\sharp$ is the weighted pseudo-inverse of $\boldsymbol{J}_P$.

Optimization

Solutions of Eq.(7.1) represents all the feasible combination of $\ddot{\boldsymbol{\theta}}_G, \boldsymbol{\tau}_C$ and $\boldsymbol{\tau}_J$. The task of the optimization part is to find the optimal solution of Eq.(7.1) where the generalized accelerations become as close as possible to the desired accelerations. The optimized accelerations are integrated to derive the joint angle data.

First, we derive the weighted least-square solution of Eq.(7.2) and the null space of $\boldsymbol{W}$ regardless of the desired acceleration by

$$\boldsymbol{x} = \boldsymbol{W}^\sharp\boldsymbol{u} + (\boldsymbol{E} - \boldsymbol{W}^\sharp\boldsymbol{W})\boldsymbol{y} \tag{7.10}$$

where $\boldsymbol{W}^\sharp$ is the pseudo inverse of $\boldsymbol{W}$, $\boldsymbol{y}$ is an arbitrary vector, and $\boldsymbol{E}$ is the identity matrix of the appropriate size. Taking the upper rows of Eq.(7.10) corresponding to the generalized accelerations, we obtain

$$\ddot{\boldsymbol{\theta}}_G = \ddot{\boldsymbol{\theta}}_G^0 + \boldsymbol{V}_G\boldsymbol{y} \tag{7.11}$$

where $\ddot{\boldsymbol{\theta}}_G^0$ is the generalized acceleration in the least-square solution of Eq.(7.10).

Next, we determine the arbitrary vector $\boldsymbol{y}$ to minimize the acceleration error by

$$\boldsymbol{y} = \boldsymbol{V}_G^*(\ddot{\boldsymbol{\theta}}_G^d - \ddot{\boldsymbol{\theta}}_G^0) \tag{7.12}$$

where $\boldsymbol{V}_G^*$ is the singularity-robust inverse [64] of $\boldsymbol{V}_G$.

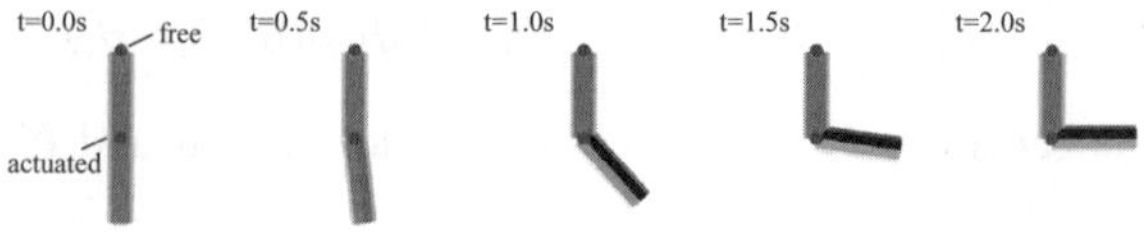

Fig. 7.2. The reference motion for the 2DOF arm.

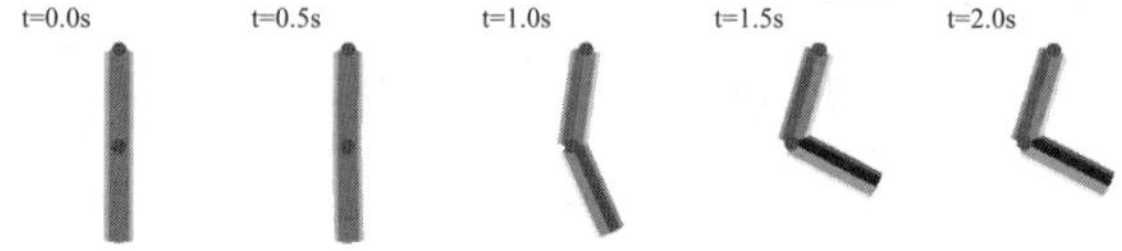

Fig. 7.3. The output of the dynamics filter for the 2DOF arm example.

Finally, substituting $\boldsymbol{y}$ into Eq.(7.10), we get the optimized solution of $\boldsymbol{x}$. Because $\boldsymbol{x}$ includes the generalized acceleration, joint torques, and constraint forces, the optimization part plays three roles at the same time: (1) computation of optimized motion, (2) computation of joint torques to realize the computed acceleration, and (3) dynamics simulation of the result.

7.3.4 Applications

We used a 28 DOF skeleton model (7 DOF legs, 4 DOF arms, a 3 DOF waist joint, and a 3 DOF neck joint) in the following examples except for the 2 DOF arm in the first example. In all the examples, the physically consistent accelerations were computed every 2 ms. The current implementation of the filter takes 70 to 80 ms per frame on an Alpha 21264 500MHz processor for the 28 DOF model.

The additional control point $\boldsymbol{P}$ was taken at the neck and its position and orientation were computed off-line, although on-line computation could be realized by an easy improvement of the implementation. In the examples with motion capture data, the original data was taken 60 frames per second and interpolated to be used as the reference joint position, velocity and accelerations.

Video clips of the motions are available online at
http://www.ynl.t.u-tokyo.ac.jp/publications/archives/video/.

2 DOF Arm

We also applied the dynamics filter to a simple 2DOF underactuated arm to demonstrate the function of the dynamics filter. The root joint of the arm is free and the other joint is actuated. We selected this mechanism because the dynamics filter is effective for underactuated mechanisms whose motion may not be dynamically feasible with any choice of actuator inputs. We used the reference motion in Fig. 7.2

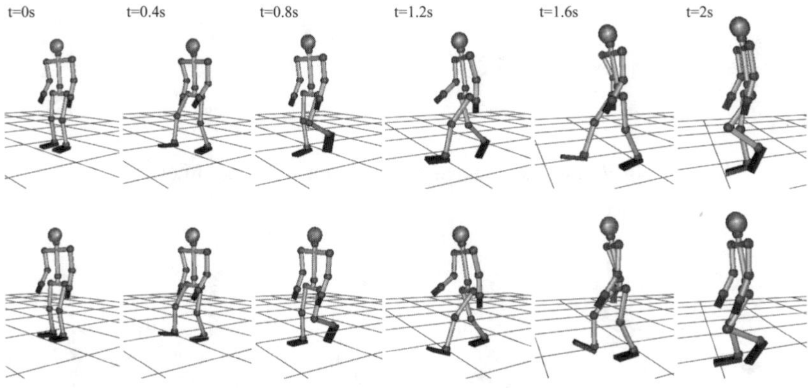

Fig. 7.4. Captured (above) and filtered (below) walking motions.

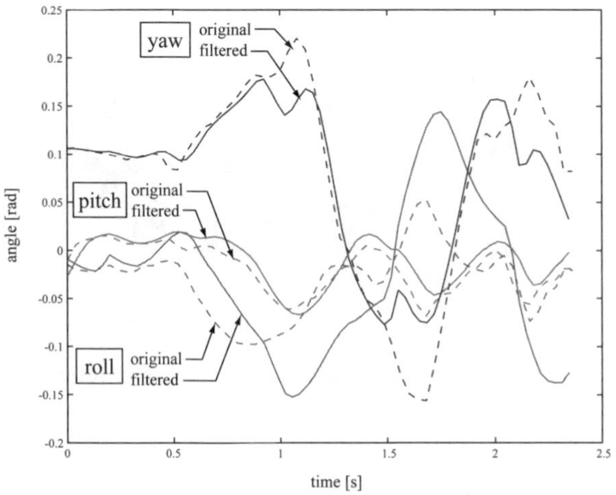

Fig. 7.5. Numerical comparison of roll, pitch, and yaw angles of the body.

and obtained the output shown in Fig. 7.3. The reference motion was physically inconsistent because of the free root joint, but it was corrected by the dynamics filter by changing the trajectory of the root joint.

Filtering Raw Motion Capture Data

Fig. 7.4 compares the captured (above) and filtered (below) walking motions. The roll, pitch, and yaw angles of the body of the original and filtered motions are compared in Fig. 7.5.

This method is applicable to any motion as shown in Fig. 7.6, which implies that we do not need to prepare different filters for different motions.

Fig. 7.6. Karate kick generated by the dynamics filter.

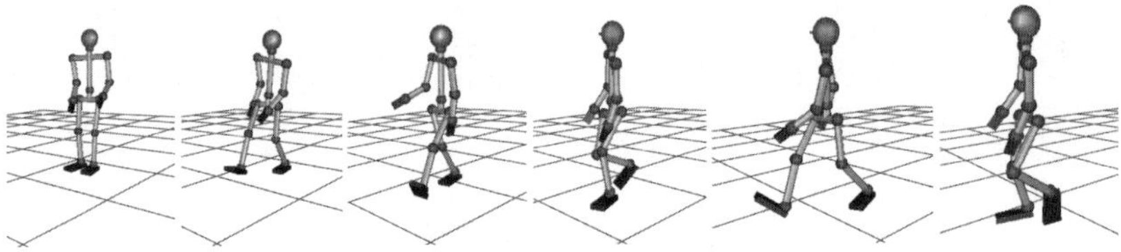

Fig. 7.7. Walk on a down slope.

Fig. 7.8. Motion generated from a kinematically combined reference motion.

Filtering into a Different Environment

We applied the same walking motion as in the previous example to a down slope. The result is shown in Fig. 7.7. No modification on captured data was made except for the position of the body and the neck, which was modified to maintain the same height from the ground as in the original.

Filtering Kinematically Synthesized Motion

The dynamics filter accepts not only raw captured data but also kinematically synthesized motions. A walking motion with a thirty-degrees turn was generated (Fig. 7.8) from the reference motion obtained by smoothly connecting two walking motions heading different directions. The reference motion does not take into account any dynamic effect such as centrifugal forces.

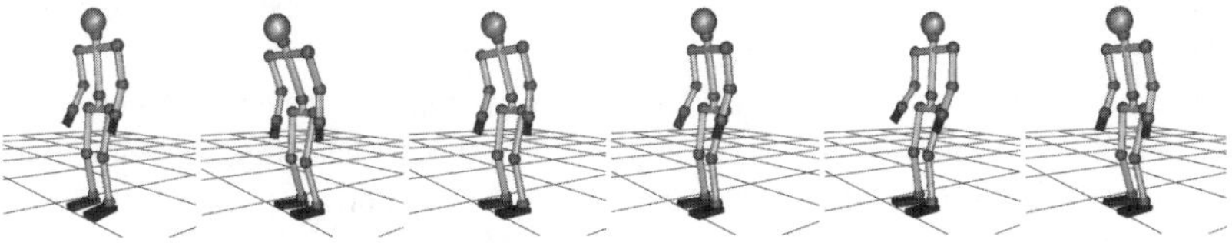

Fig. 7.9. Example of interactive motion generation: push a standing figure.

Interactive Motion Generation

Fig. 7.9 demonstrates the interactivity of our approach, where the figure is controlled to keep standing by the dynamics filter and reacts to the force applied by the user.

7.3.5 Discussion

The original and filtered walking motions showed good correspondence in the motions presented here, although we observed small latencies in some degrees of freedom such as the roll angle of the body (Fig. 7.5). It turned out that the actual accelerations of those degrees of freedom were much more different from the desired ones than the others, probably because they had to be modified by the optimizer to compensate for the physical consistency or the difference in the environment. Their angles will not follow the original trajectory until the feedback terms in Eq.(7.6) become large enough. In fact, the width of the motion of the roll angle of the body is much larger in the filtered motion than in the original one.

The examples in Figs. 7.7 and 7.8 showed the possibility of applying the dynamics filter to adapt motion capture data to the environment different from the one where the motion was actually performed, and to synthesize physically consistent motions from a motion capture data library. In these examples, we could generate valuable motions by quite simple kinematic modifications of the original walking motion. For more practical applications, however, careful kinematic preprocessing would be required to adjust the feet placements and constraints.

The problem of the dynamics filter is that it is very difficult to tune the parameters (feedback gains, weights for weighted pseudo inverses) and we need to find a set of parameters for each behavior. We will then encounter a problem in generating intermediate motions just as in using task-specified controllers. The encouraging fact, however, is that it is rather easier to interpolate the parameters than to interpolate the controllers to get intermediate behaviors.

Unfortunately, we could not find a systematic method for tuning the parameters. If the reference motion is physically consistent, larger gains would work because the figure only has to follow the reference motion as precisely as possible. Otherwise, we have to select the appropriate gains according to the importance of the trajectory of each joint in order not to overfit the joints to the original trajectory and to allow flexible interactions. We would also have to use smaller weights for less important joints so that the SR-inverse can allow enough deviation from the desired accelerations to compensate for the physical inconsistency.

The main contribution of the dynamics filter is that it provides a general framework for on-line generation of physically consistent motions using an efficient algorithm for dynamics simulation of human figures. In our current implementation, it is not guaranteed that the resulting motion would always become close to the reference. The result will be far from the original if we give extremely inconsistent motions as reference. A possible solution to this problem is to apply some task-specific controllers so that the figure will not fall down. For the standing human figure in Fig. 7.9, for example, we could design a controller that allows the human figure to step out if the force is too large to resist. Designing controllers for robotics systems is a common problem in various field in robotics, and a number of controllers are proposed. In the current implementation of the dynamics filter, we adopted a simple feedback controller consisting of a feedback loop in joint space and another in the Cartesian space of a selected link. The results presented in this chapter proved that, even with the simple controllers we used, the dynamics filter can generate physically consistent motions on line using reference motions from various sources. It should also be possible to apply more sophisticated controllers proposed by many researchers, which will be included in our future work.

7.4 Summary

The contributions of this chapter are summarized as follows:

(1) The concept of a dynamics filter is proposed. A dynamics filter is a motion generator that creates a physically consistent motion from any reference motion that may be physically infeasible for the model.
(2) A method for simulating collisions and contact was presented. The method is based on our previous work on structure-varying kinematic chains, which was extended to handle unilateral constraints by a fast trial-and-error procedure.
(3) The implementation of an on-line dynamics filter was described and proved the potential ability of the dynamics filter in interactive motion generation by a number of motions generated from motion capture data.

8

Synergetic Choreography of Human Figures

8.1 Introduction

This chapter presents an intuitive interface called *pin-and-drag interface* for generating whole-body motions of human figures considering only kinematic constraints. Users can choose any number of *pinned* links whose positions and/or orientations are fixed and then specify the position of a *dragged* link through any intuitive interface such as a mouse or a joystick. The interface is based on an inverse kinematics algorithm called *UTPoser*, which computes a numerically stable posture that brings the dragged link to the specified position while satisfying various constraints including pinned links, joint motion ranges and desired joint angles. The method is capable of creating human-like motions without any reference motion, although editing prerecorded motion is also one of its possible applications.

UTPoser has two major features. The first is that it can handle various types of constraints at the same time by solving instantaneous kinematics equations on joint velocities. Current implementation includes constraints on position of pinned and dragged links, joint motion ranges, and desired joint values, although we could add other constraints that can be written as a set of equations in joint velocities. The second feature of the method is that it yields reasonable results even if the human figure is in kinematic singularity or the constraints are inconsistent. This feature is realized by applying singularity-robust (SR) inverse [64] to solve the kinematics equation. We can therefore add any number of constraints at will without making the results unstable. Given any set of constraints, the inverse kinematics computation will yield a reasonable result that satisfies the constraints as much as possible.

In spite of the simple interface and computation, the generated motions look quite *natural* as a human motion, although its qualitative evaluation is difficult and

---

This chapter was adapted from, by permission, K. Yamane and Y. Nakamura, "Synergetic CG Choreography through Constraining and Deconstraining at Will," Proceedings of IEEE International Conference on Robotics and Automation, pp.855–862, Washington DC, May 2002, and K. Yamane and Y. Nakamura, "Natural Motion Animation through Constraining and Deconstraining at Will," vol.9, no.3, pp.352–360, 2003.

not the scope of this book. The reason for the natural-looking output is not clear. However, it is possible that *synergetic* effects act an important role in determining the whole-body motion from relatively few inputs.

This chapter is organized as follows. In Section 8.2, the overview of the method is described along with introductions on mathematical preliminaries and relationships with previous work. The computational details of the inverse kinematics are described in Section 8.3. Section 8.4 describes how to extend the method to *motion editing in motion*, online modification of a pre-recorded motion. Section 8.5 shows example motions generated by this method.

8.2 Overview

8.2.1 Pin-and-Drag Interface

The task of the computational engine for the pin-and-drag interface is to generate a motion in which

(1) the link specified by the user (the dragged link) follows the indicated path,
(2) any number of links specified by the user (pinned links) stay at their reference positions and/or orientations,
(3) each joint angle stays in its motion range, and
(4) each joint angle stays as close as possible to the given reference angle.

There are two obvious difficulties in computing the solution that satisfies all of these constraints:

- it is difficult (or virtually impossible) to derive an analytical method that can handle general cases, and
- the constraints often conflict with each other (consider the case where the user drags a link beyond the reachable space determined by the pinned links)

The first problem comes from the fact that the constraints are expressed by a set of complicated nonlinear equations, and the second implies that these equations may not have an exact solution.

The first problem is solved by introducing differential kinematics that give a linear relationship between the constraints and the joint velocities. In order to deal with the second problem, we divide the four constraints into two priority levels [65]. The first constraint (the dragged link) is given the higher priority and is always satisfied exactly, while the other constraints are given the lower priority. The null-space of the first constraint is used to satisfy the constraints with the lower priority. If there is a conflict among the constraints, the least-square optimization is applied to find the best approximation for the lower-priority constraints.

Although the null-space decomposition and the least-square solution are commonly done with a pseudoinverse, it may result in extremely large and, therefore, physically infeasible solutions in the neighborhood of singularity. The singularity-robust (SR) inverse [64] is adopted to avoid this problem. The SR inverse eases the singularity problem by allowing errors around singular points. We introduce the feedback controller as a device for the recovery of errors which the singularities or conflicts introduce. By integrating the SR inverse and the feedback controller into the differential kinematics of constrained kinematic chains, the pin-and-drag interface is equipped with the "elastic" property, the natural response, and reliability.

8.2.2 Differential Kinematics with Redundancy

The Jacobian matrix of the position of a link with respect to the joint angles is defined as:

$$\boldsymbol{J}_i \triangleq \frac{\partial \boldsymbol{r}_i}{\partial \boldsymbol{\theta}} \tag{8.1}$$

where $\boldsymbol{r}_i$ is the position of link i, $\boldsymbol{\theta}$ is the vector composed of all joint angles, and $\boldsymbol{J}_i$ is the Jacobian matrix of $\boldsymbol{r}_i$ with respect to $\boldsymbol{\theta}$. An efficient method for computing the Jacobian matrix can be found in [72]. The velocity of link i and joint angles are related as

$$\dot{\boldsymbol{r}}_i = \boldsymbol{J}_i \dot{\boldsymbol{\theta}}. \tag{8.2}$$

If the base link is not fixed to the inertial frame, as is often the case with human figures, its linear and angular velocities are also included in $\dot{\boldsymbol{\theta}}$. If $\boldsymbol{J}_i$ is square and non-singular, it can be inverted to yield

$$\dot{\boldsymbol{\theta}} = \boldsymbol{J}_i^{-1} \dot{\boldsymbol{r}}_i, \tag{8.3}$$

by which we can control the joints to follow the reference trajectory of $\boldsymbol{r}_i$.

Unfortunately $\boldsymbol{J}_i$ is not square in our problem because human and animal characters typically have over 30 DOF. The general solution of Eq.(8.2) is described using the pseudoinverse $\boldsymbol{J}_i^\sharp$ as

$$\dot{\boldsymbol{\theta}} = \boldsymbol{J}_i^\sharp \dot{\boldsymbol{r}}_i + (\boldsymbol{E} - \boldsymbol{J}_i^\sharp \boldsymbol{J}_i)\boldsymbol{y} \tag{8.4}$$

where $\boldsymbol{E}$ is the identity matrix and $\boldsymbol{y}$ is an arbitrary vector. The second term shows the redundancy and reserves the degrees of freedom which can be used to find the optimal solution that accomplish other tasks without breaking Eq.(8.2) [65].

8.2.3 The Singularity-Robust Inverse

Singularity-robust (SR) inverse [64] is also known as damped pseudoinverse [54]. Consider a linear equation

$$\boldsymbol{A}\boldsymbol{x} = \boldsymbol{b}. \tag{8.5}$$

If the coefficient matrix $\boldsymbol{A}$ is not square, we usually use its pseudoinverse $\boldsymbol{A}^\sharp$ to compute the least-square solution with the minimal norm. However, the pseudoinverse solution tends to have extremely large amplitude in the neighborhood of singular points. This problem happens because the pseudoinverse minimizes the norm of the error $|\boldsymbol{b} - \boldsymbol{A}\boldsymbol{x}|$ first, and then minimizes the norm of the solution $|\boldsymbol{x}|$ [64]. The SR inverse, on the other hand, avoids this problem by minimizing the sum of the norms of the error and the solution.

For an m-by-n ($m < n$) matrix $\boldsymbol{A}$, its pseudoinverse is computed by

$$\boldsymbol{A}^{\sharp} = \boldsymbol{A}^T(\boldsymbol{A}\boldsymbol{A}^T)^{-1}. \tag{8.6}$$

$\boldsymbol{A}^{\sharp}$ may have extremely large elements when $\boldsymbol{A}\boldsymbol{A}^T$ is nearly singular. The SR inverse, on the other hand, uses the following equation instead of Eq.(8.6):

$$\boldsymbol{A}^{*} = \boldsymbol{A}^T(\boldsymbol{A}\boldsymbol{A}^T + k\boldsymbol{I})^{-1} \tag{8.7}$$

where $\boldsymbol{A}^*$ is the SR inverse of $\boldsymbol{A}$, $\boldsymbol{I}$ is an identity matrix, and k is the parameter that determines the weighting between the norm of the solution and the error. If we use small k, then the error gets small but the solution might get large around singular points, and vice versa [62].

8.2.4 The Algorithm

The algorithm to be proposed in this chapter consists of the following five steps:

(1) compute the general solutions of joint velocities that move the dragged link towards the indicated position (Section 8.3.1)
(2) compute the desired velocities of the other constraint variables, taking account of their reference and current values (Section 8.3.4)
(3) compute the Jacobian matrix of the constraint variables with respect to the joint angles (Section 8.3.3)
(4) using the general solutions in step 1, find a particular solution that closely satisfies the desired velocities of the constraint variables (Section 8.3.2)
(5) numerically integrate the joint velocities to get the joint angles.

The proposed algorithm has a number of advantages over the previous ones with similar objectives:

- moving a single link determines the posture of the whole body
- any link can be dragged or pinned
- no limit on the number of pinned links
- constraint variables can be instantly included or removed
- relative importance of the constraint variables can be tuned.

8.2.5 Comparison with Previous Work

The main objective of this chapter is to develop an interface that enables people to generate whole-body motions of articulated figures with little effort and preferably without captured data. Although there are many related work in this field, most of these efforts aim to solve the problem of editing or retargetting pre-recorded data. The authors would think that it is because the previous researchers viewed generating new motions too challenging.

Inverse kinematics is the key in our approach. Many previous work in inverse kinematics used global optimization over the spacetime constraint of motion [28, 88, 87, 47]. The SR-inverse was also employed in [28] to avoid singularities in the Jacobian matrix. Online computation using local optimization, on the other hand, was investigated by Choi *et al.* [17] based on the feedback control and the null-space method similar to ours. In [17] the pinned links are placed only at the end-links, due to the fact that the increase of constraints makes the Jacobian matrix ill-conditioned and the troubles of singularity cannot be avoided by the use of pseudoinverses. Our approach allows as many pins as we need, even at intermediate links or two neighboring links, thanks to the SR-inverse.

Badler *et al.* [5] and Phillips *et al.* [80] also developed a pin-and-drag interface, and implemented it as a part of the 3D animation system *Jack* [79]. Our formulation follows the previous work in principle and includes the following improvements:

- In Badler's system, the link hierarchy was recomputed so that the dragged link becomes the root, while in ours the link hierarchy does not change once the structure is given, thanks to the virtual link representation of closed loops presented in Chapter 3. We can eliminate the overhead to switch the dragged links, providing more responsive and comfortable interface to the user.
- Badler's system considered joint motion range of rotational (or 1DOF) joints, and it would find a difficulty to include spherical-joint limits with their projection scheme.
- The normal pseudoinverse was used in their computation of inverse kinematics. The SR-inverse we use, in contrast, allows us to apply as many constraints as we need, without concern about the singularity.

The strict comparison of computational cost between the proposed and the previous methods is not straightforward. One difference in computational expense comes from the elimination of link-hierarchy recomputation. Another difference is due to the SR-inverse as opposed to the pseudoinverse in the previous work. In [64] the computational comparison was extensively made between the pseudoinverse and the SR-inverse and the difference was negligibly small for full-rank matrices. For non-full-rank matrices the SR-inverse is known to be significantly more efficient than the pseudoinverse. These algorithmic differences show the computational advantage of the proposed method in this chapter.

Representation of the motion range of 3DOF spherical joints is fundamental to obtaining natural behaviors of human characters. A simple inequality representation of the Euler angles is inappropriate due to their nonlinearity [6, 56]. A precise anatomical modeling was recently proposed [56], where three 3DOF spherical joints and a 5DOF joint were used for modeling. It is common in the previous work that the 3DOF of a spherical joint are parameterized by three variables known as the Euler angles and, therefore, suffers from the well-known algorithmic singularity. In contrast, we propose to parameterize the 3DOF of a spherical joint in a unique definition of three variables. It has a singular point only at a physically insignificant direction. In addition, when the limb configuration goes beyond the motion range, a feedback control is applied to force the range. The feedback control is proposed in a simpler form in this chapter than those used in the previous work.

8.3 Computational Details

8.3.1 The Dragged Link

First we compute $\dot{\boldsymbol{\theta}}$ with which the dragged link exactly follows its reference velocity $\dot{\boldsymbol{r}}_P^{ref}$ and position $\boldsymbol{r}_P^{ref}$. Let $\boldsymbol{r}_P$ denote the current position of the dragged link. Its desired velocity is computed by

$$\dot{\boldsymbol{r}}_P^d = \dot{\boldsymbol{r}}_P^{ref} + \boldsymbol{K}_P(\boldsymbol{r}_P^{ref} - \boldsymbol{r}_P) \tag{8.8}$$

where $\boldsymbol{K}_P$ is a positive-definite gain matrix. The relationship between $\dot{\boldsymbol{\theta}}$ and $\dot{\boldsymbol{r}}_P$ is given by

$$\dot{\boldsymbol{r}}_P = \boldsymbol{J}_P\dot{\boldsymbol{\theta}} \tag{8.9}$$

where $\boldsymbol{J}_P$ is the Jacobian matrix of $\boldsymbol{r}_P$ with respect to the joint angles. The general solution $\dot{\boldsymbol{\theta}}$ for the desired velocity $\dot{\boldsymbol{r}}_P^d$ is computed by

$$\dot{\boldsymbol{\theta}} = \boldsymbol{J}_P^{\sharp}\dot{\boldsymbol{r}}_P^d + (\boldsymbol{E} - \boldsymbol{J}_P^{\sharp}\boldsymbol{J}_P)\boldsymbol{y}. \tag{8.10}$$

The feedback control is applied only to compensate the numerical errors. A weighted pseudoinverse [62] may be used instead of the normal pseudoinverse to characterize the joint motions.

8.3.2 Lower-Priority Constraints

The general solutions of Eq.(8.10) is rewritten as

$$\dot{\boldsymbol{\theta}} = \dot{\boldsymbol{\theta}}_0 + \boldsymbol{W}\boldsymbol{y} \tag{8.11}$$

where $\boldsymbol{W} \triangleq \boldsymbol{E} - \boldsymbol{J}_P^{\sharp}\boldsymbol{J}_P$ and $\dot{\boldsymbol{\theta}}_0 \triangleq \boldsymbol{J}^{\sharp}\dot{\boldsymbol{r}}_P^d$.

Suppose we have N_F pinned links whose positions are denoted by $\boldsymbol{r}_{Fi}(i = 1 \dots N_F)$, N_D joints with their reference angles $\boldsymbol{\theta}_D$, and N_L joints with their joint values $\boldsymbol{\theta}_L$ out of the motion ranges. Note that N_L may vary anytime during the motion, whereas N_D stays constant until it is changed by the higher level of control. Using the vectors, we define $\boldsymbol{p}_{aux}$ as follows:

$$\boldsymbol{p}_{aux} \triangleq \left(\boldsymbol{r}_{F1}^T \dots \boldsymbol{r}_{FN_F}^T\ \boldsymbol{\theta}_D^T\ \boldsymbol{\theta}_L^T\right)^T \tag{8.12}$$

whose velocity $\dot{\boldsymbol{p}}_{aux}$ is related to the joint velocity $\dot{\boldsymbol{\theta}}$ by a relationship similar to Eq.(8.2):

$$\dot{\boldsymbol{p}}_{aux} = \boldsymbol{J}_{aux}\dot{\boldsymbol{\theta}}. \tag{8.13}$$

Computation of $\boldsymbol{J}_{aux}$ is to be discussed in the following section.

The arbitrary vector $\boldsymbol{y}$ is computed as follows. We first compute the desired velocity $\dot{\boldsymbol{p}}^d_{aux}$ of $\boldsymbol{p}_{aux}$ to take account of the errors between the constraint conditions and their current values as described in Section 8.3.4. Substituting Eq.(8.11) into Eq.(8.13) yields

$$\dot{\boldsymbol{p}}_{aux} = \dot{\boldsymbol{p}}^0_{aux} + \boldsymbol{J}_{aux}\boldsymbol{W}\boldsymbol{y} \tag{8.14}$$

where $\dot{\boldsymbol{p}}^0_{aux} \triangleq \boldsymbol{J}_{aux}\dot{\boldsymbol{\theta}}_0$. Using $\boldsymbol{S} \triangleq \boldsymbol{J}_{aux}\boldsymbol{W}$ and $\Delta\dot{\boldsymbol{p}}_{aux} \triangleq \dot{\boldsymbol{p}}^d_{aux} - \dot{\boldsymbol{p}}^0_{aux}$, we have a simpler form of the equation:

$$\boldsymbol{S}\boldsymbol{y} = \Delta\dot{\boldsymbol{p}}_{aux}. \tag{8.15}$$

Since $\boldsymbol{S}$ is not always well conditioned, we use the SR inverse to solve this problem. Denoting the SR inverse of $\boldsymbol{S}$ by $\boldsymbol{S}^*$, $\boldsymbol{y}$ is computed by

$$\boldsymbol{y} = \boldsymbol{S}^*\Delta\dot{\boldsymbol{p}}_{aux}. \tag{8.16}$$

The joint velocity $\dot{\boldsymbol{\theta}}$ is obtained by substituting Eq.(8.16) into Eq.(8.11), which is then integrated to yield the joint angle $\boldsymbol{\theta}$ for animation.

8.3.3 Computation of $\boldsymbol{J}_{aux}$

Let $\boldsymbol{J}_{Fi}(i = 1 \ldots N_F)$ be the Jacobian matrix of $\boldsymbol{r}_{Fi}$ with respect to the joint angles. Then, for all pinned links we have

$$\dot{\boldsymbol{r}}_{Fi} = \boldsymbol{J}_{Fi}\dot{\boldsymbol{\theta}}. \tag{8.17}$$

For the joints with reference angles, the relationship between their velocities $\dot{\boldsymbol{\theta}}_D$ and $\dot{\boldsymbol{\theta}}$ is described by

$$\dot{\boldsymbol{\theta}}_D = \boldsymbol{J}_D\dot{\boldsymbol{\theta}} \tag{8.18}$$

where $\boldsymbol{J}_D$ is the matrix whose (i, j)-th element is 1 if the i-th element of $\boldsymbol{\theta}_D$ corresponds to the j-th element of $\boldsymbol{\theta}$, and 0 otherwise.

Similarly, we can describe the relationship between $\dot{\boldsymbol{\theta}}$ and the velocity of $\boldsymbol{\theta}_L$ as follows:

$$\dot{\boldsymbol{\theta}}_L = \boldsymbol{J}_L\dot{\boldsymbol{\theta}} \tag{8.19}$$

where $\boldsymbol{J}_L$ is the matrix whose (i, j)-th element is 1 if the i-th element of $\boldsymbol{\theta}_L$ corresponds to the j-th element of $\boldsymbol{\theta}$, and 0 otherwise.

Combining the above-defined matrices, $\boldsymbol{J}_{aux}$ is formed as follows:

$$\boldsymbol{J}_{aux} = \left(\boldsymbol{J}^T_{F1} \ldots \boldsymbol{J}^T_{FN_F}\ \boldsymbol{J}^T_D\ \boldsymbol{J}^T_L\right)^T. \tag{8.20}$$

The computation of columns of $\boldsymbol{J}_{Fi}, \boldsymbol{J}_P$ and $\boldsymbol{J}_L$ corresponding to spherical joints is to be discussed in Section 8.3.5.

8.3.4 Computation of $\dot{\boldsymbol{p}}^d_{aux}$

The desired velocity of each pinned link $\dot{\boldsymbol{r}}^d_{Fi}$ is computed by the following feedback law:

$$\dot{\boldsymbol{r}}^d_{Fi} = \boldsymbol{K}_{Fi}(\boldsymbol{r}^{ref}_{Fi} - \boldsymbol{r}_{Fi}) \tag{8.21}$$

where $\boldsymbol{r}^{ref}_{Fi}$ is the reference position, and $\boldsymbol{K}_{Fi}$ is a positive-definite gain matrix.

The desired velocity of joints with their reference angles, $\dot{\boldsymbol{\theta}}^d_D$, is computed by

$$\dot{\boldsymbol{\theta}}^d_D = \boldsymbol{K}_D(\boldsymbol{\theta}^{ref}_D - \boldsymbol{\theta}_D) \tag{8.22}$$

where $\boldsymbol{\theta}^{ref}_D$ represents the reference joint angles, and $\boldsymbol{K}_D$ is a positive-definite gain matrix.

The desired velocities of joints that exceed their motion ranges are computed as follows:

$$\dot{\theta}^d_{Li} = \begin{cases} K_{Li}(\theta^{max}_{Li} - \theta_{Li}) \text{ if } (\theta_{Li} > \theta^{max}_{Li}) \\ K_{Li}(\theta^{min}_{Li} - \theta_{Li}) \text{ if } (\theta_{Li} < \theta^{min}_{Li}) \end{cases} \tag{8.23}$$

where θ^{max}_{Li} and θ^{min}_{Li} are the maximum and minimum joint angles, respectively, and K_{Li} is a positive scalar gain.

Equations (8.22) and (8.23) work for 1DOF joints. The following subsection extends the ideas to 3DOF spherical joints.

8.3.5 Handling Spherical Joints

Reference Joint Displacements

The joint displacement and the joint velocity of a spherical joint are represented by the 3×3 orientation matrix $\boldsymbol{R}_i$ and its associated angular velocity $\boldsymbol{\omega}_i$ respectively, described in its parent link frame.

When a spherical joint is given a reference joint displacement $\boldsymbol{R}_{Di} \in \boldsymbol{R}^{3\times3}$, we compute its desired joint velocity as follows. We first compute the error vector $\boldsymbol{e}_i$ between the current joint displacement $\boldsymbol{R}_i$ and $\boldsymbol{R}_{Di}$ by

$$\boldsymbol{e}_i = \frac{1}{2}\begin{pmatrix} \Delta\boldsymbol{R}_i(1,2) - \Delta\boldsymbol{R}_i(2,3) \\ \Delta\boldsymbol{R}_i(1,3) - \Delta\boldsymbol{R}_i(3,1) \\ \Delta\boldsymbol{R}_i(2,1) - \Delta\boldsymbol{R}_i(3,2) \end{pmatrix} \tag{8.24}$$

$$\Delta\boldsymbol{R}_i \triangleq \boldsymbol{R}_{Di}\boldsymbol{R}^T_i \tag{8.25}$$

where $\Delta\boldsymbol{R}_i(m,n)$ denotes the (m,n)-th element of $\Delta\boldsymbol{R}_i$. Then, the desired angular velocity $\boldsymbol{\omega}^d_{Di}$ is computed by [53]

$$\boldsymbol{\omega}^d_{Di} = -\boldsymbol{K}_{Di}\boldsymbol{e}_i \tag{8.26}$$

where $\boldsymbol{K}_{Di}$ is a positive-definite gain matrix. Equations (8.24)–(8.26) are used for spherical joints in place of Eq.(8.22).

Also included in $\boldsymbol{J}_{Fi}, \boldsymbol{J}_D$ and $\boldsymbol{J}_L$ corresponding to a spherical joint are three columns associated with the angular velocity. Each column is computed just as if there is a rotational joint around x, y or z axis.

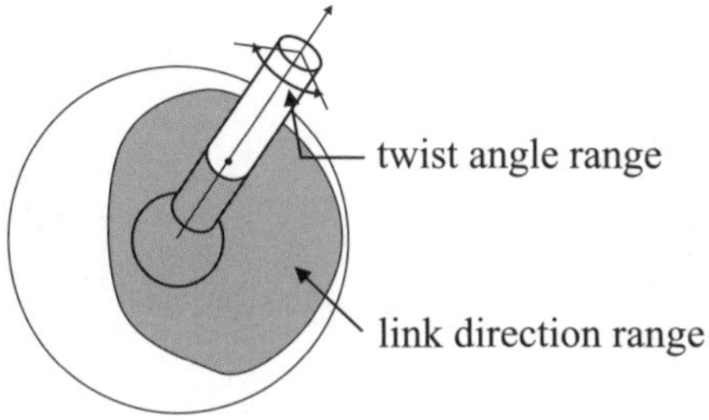

Fig. 8.1. Joint motion range of a spherical joint.

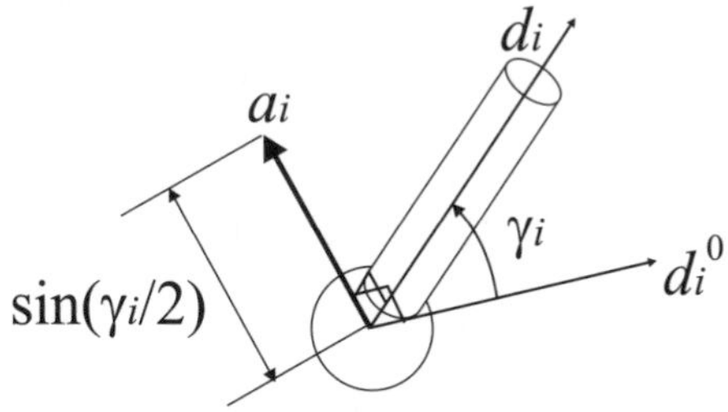

Fig. 8.2. Graphical representation of $\boldsymbol{a}_i$ and γ_i.

Joint Motion Range

The motion range of a spherical joint is expressed as a region in a three dimensional space. Simplicity of the geometric representation of the region is important for realtime computation. The region would show a complex shape if we represent it with common coordinates such as the Euler angle due to their nonlinearity. In this subsection, we propose an intuitive representation of spherical joint motion range. Although one may see a similarity to the *equivalent angle-axis* representation [18], it is different in the sense that our representation consists of two sequential rotations. The representation provides three parameters, two of which describe the link direction and the other denotes the twist angle, as illustrated in Fig. 8.1.

When $\boldsymbol{R}_i$ is the identity, the link is at the nominal direction and we represent it by unit vector $\boldsymbol{d}_i^0$. The current link direction $\boldsymbol{d}_i$ is obtained by rotating $\boldsymbol{d}_i^0$ about vector $\boldsymbol{a}_i$ that lies in the two-dimensional plane orthogonal to $\boldsymbol{d}_i^0$. The magnitude of $\boldsymbol{a}_i$ is not 1, but $\sin(\gamma_i/2)$ where γ_i represents the angle of rotation, as seen in Fig. 8.2. The twist angle α is defined as the angle by which the link frame after the first rotation around $\boldsymbol{a}_i$, is rotated to make the current link frame $\boldsymbol{R}_i$. The entire configuration of a spherical joint is therefore included in a cylinder whose axis is $\boldsymbol{d}_i^0$, radius 1, and height 2π.

In our implementation, $\boldsymbol{d}_i^0$ is set as $(1\ 0\ 0)^T$ for all joints and therefore $\boldsymbol{a}_i$ stays in the yz plane, namely $\boldsymbol{a}_i = (0\ a_y\ a_z)^T$. Thus the motion range is described by a cylinder with an axis parallel to the α axis as shown in Fig. 8.3.

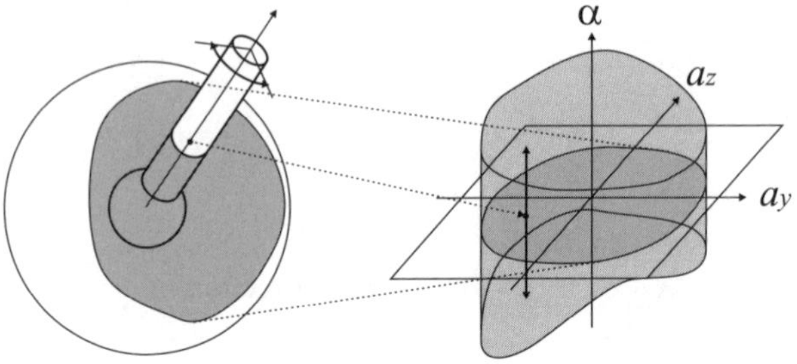

Fig. 8.3. Motion range of a spherical joint projected onto (a_y, a_z, α) space.

a_y, a_z and α are computed as follows: Since $\boldsymbol{d}_i^0 = (1\ 0\ 0)^T$, we have

$$\begin{aligned} \boldsymbol{d}_i &= \boldsymbol{R}_i \boldsymbol{d}_i^0 \\ &= (\boldsymbol{R}_i(1,1)\ \boldsymbol{R}_i(2,1)\ \boldsymbol{R}_i(3,1))^T. \end{aligned} \tag{8.27}$$

Therefore a_y and a_z are computed by

$$a_y = -\frac{\boldsymbol{R}_i(3,1)}{\sqrt{2(1+\boldsymbol{R}_i(1,1))}} \tag{8.28}$$

$$a_z = \frac{\boldsymbol{R}_i(2,1)}{\sqrt{2(1+\boldsymbol{R}_i(1,1))}}. \tag{8.29}$$

Then, α is computed by comparing the y and z axes of the frame after the rotation around $\boldsymbol{a}_i$ and the actual current frame since their x axes coincide with each other. Although Eqs.(8.28)(8.29) show singularity at $\gamma_i = \pm\pi$, it is not a practical problem since these values are usually beyond joint limits.

Our next step is to determine whether the computed parameters are inside or outside of the motion range. For ease of computation, we describe the link direction range by a collection of triangle patches in a_y-a_z plane. The whole motion range is represented by a collection of triangular cylinders with the triangle patches as their footprints and α axis as their axes.

We first look for the triangle in which $(a_y, a_z, 0)$ is included. If no triangle is found, the joint is out of the joint motion range. Otherwise, we then proceed to check if (a_y, a_z, α) is between the upper and lower limits of the triangle.

If the parameters (a_y, a_z, α) are out of the range, we compute the desired velocity to bring the joint back into the range and include it in $\dot{\boldsymbol{p}}^d_{aux}$. For this purpose, we define the standard orientation $\boldsymbol{R}_{Si}$ for each joint, and compute the desired joint velocity $\boldsymbol{\omega}_{Li}$ to move the joint towards $\boldsymbol{R}_{Si}$. This is achieved by simply substituting $\boldsymbol{R}_{Si}$ into $\boldsymbol{R}_{Di}$ in Eq.(8.25) and $\boldsymbol{\omega}_{Li}$ into $\boldsymbol{\omega}_{Di}$ in Eq.(8.26). The more theoretical alternative would be to control the joint towards the configuration on the boundary of the motion range as in [56]. We did not take this for computational simplicity sake.

Fig. 8.4 shows the motion range of a right shoulder. The light colored (or red for readers with colored figures) shows the upper surfaces of triangular cylinders,

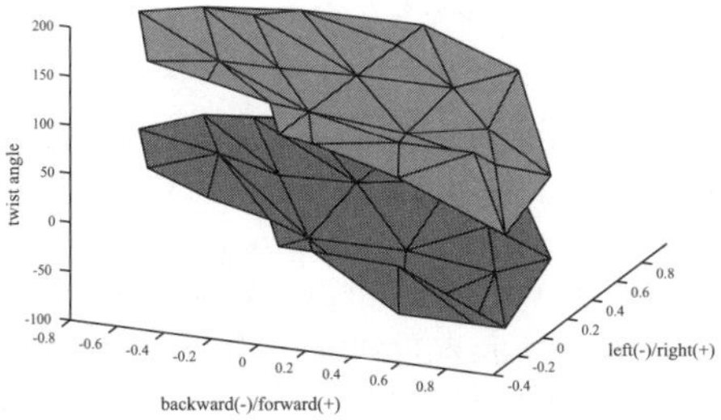

Fig. 8.4. An example of a motion range of a spherical joint.

while the dark colored (or blue) shows their lower surfaces. The vertical axis denotes the twist angle in degree, while the other two denote the rotation angle in forward/backward and left/right directions as indicated in the figure. This example includes 35 triangles in the a_y-a_z plane. We modeled the motion ranges of 10 spherical joints with 8 to 35 triangles depending on their shapes. Due to the simplicity of computation, handling spherical joints did not affect the realtime performance of the system.

8.4 Editing Motion in Motion

In Section 8.3 we discussed the algorithm for static pin-and-drag interface, where the reference positions of pins, $\boldsymbol{r}_{Fi}^{ref}$, and the reference joint angles, $\boldsymbol{\theta}_{Di}^{ref}$, were assumed constant, while the dragged link had its velocity $\dot{\boldsymbol{r}}_P^{ref}$. Extending the algorithm to include velocities of $\boldsymbol{r}_{Fi}^{ref}$ and $\boldsymbol{\theta}_{Di}^{ref}$ for dynamic pin-and-drag interface is straightforward. By this extension, we can apply the algorithms to editing motion in motion and retargetting captured data. Changing the pins and the dragged link one after another in motion enables us to generate step by step rich and complex motions of characters.

The following two slight modifications are required to extend the above method to editing motion in motion:

- The position of the pins are obtained by direct kinematics computation for each frame of the reference motion. Since each pin has reference velocity $\dot{\boldsymbol{r}}_{Fi}^{ref}$, the following equation is used instead of Eq.(8.21):

$$\dot{\boldsymbol{r}}_{Fi}^{d} = \dot{\boldsymbol{r}}_{Fi}^{ref} + \boldsymbol{K}_{Fi}(\boldsymbol{r}_{Fi}^{ref} - \boldsymbol{r}_{Fi}). \tag{8.30}$$

- The reference joint displacements are set as those of the reference motion. Using the reference joint velocities $\dot{\boldsymbol{\theta}}_D^{ref}$, the following equation is used instead of Eq.(8.22):

$$\dot{\boldsymbol{\theta}}_D^{d} = \dot{\boldsymbol{\theta}}_D^{ref} + \boldsymbol{K}_D(\boldsymbol{\theta}_D^{ref} - \boldsymbol{\theta}_D). \tag{8.31}$$

8.5 Examples

The proposed method was implemented as the computational engine of CG animation software *Animanium*^{TM} developed by Sega Corporation, where the method is used for generating key frames and recording realtime manipulation of human figures. The software is equipped with graphical interfaces to select pinned and dragged links, define the weight, feedback gain and motion range for each joint. The motion of the mouse is mapped into the three-dimensional motion of the dragged link. The users can create appealing animations by simply specifying key frames using the interface, which are then interpolated automatically to generate the animation. The computation time is approximately 33ms on PentiumIII 1GHz processor, for a 48DOF human figure with 1 dragged link, 5 pins, and 20 joints with reference joint displacements and joint motion ranges.

8.5.1 Pin and Drag

Fig. 8.5 shows various postures generated by a single pin-and-drag procedure from the initial posture (a), with both hands and feet pinned.

8.5.2 Effect of Joint Motion Range

The effect of considering joint motion range is shown in Fig. 8.6, where the feet are pinned and the head is dragged from the original posture (a) to the final (d). As the user interface for changing joint motion range, the motion ranges of spherical joints are shown by cones representing the link direction ranges in the figure. Although the neck joint exceeds its range in posture (b), which is represented by red cone, it returns back to the range by bending back the chest joint in (c). All joints stay in the range in the final posture (d). Note that this natural-looking behavior is generated in real time by a single pin-and-drag procedure taking account of the joint motion ranges.

8.5.3 Realtime Motion Generation

Images in Fig. 8.7 is taken from a video clip recorded also in realtime when the user dragged the right hand of the character for 4 seconds. The pins were set at five links — the toes, heels, and the left hand, shown in blue. Note that we can set a pin at link not necessarily at the end of a chain, such as heel links. A single drag created a realistic motion like picking up an object from the floor.

8.5.4 Editing Motion in Motion

Figs. 8.8 and 8.9 show results of editing pre-recorded motion. The motion in Fig. 8.8 was created by a professional animator using the software, and the modification was done by a naive user. The original motion was a short walk consisting of 6 keyframes.

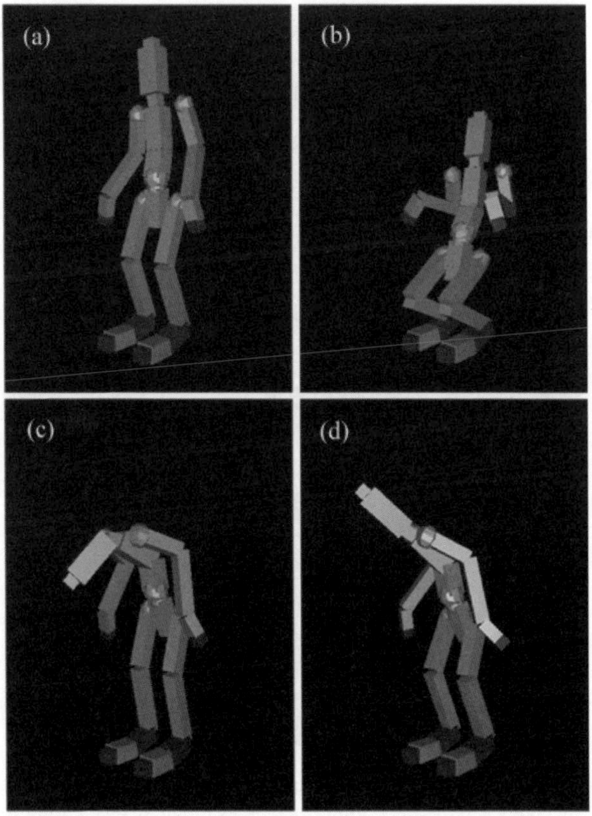

Fig. 8.5. Postures generated by a single pin-and-drag; (a) original posture, (b) pelvis dragged, (c) head dragged, (d) left shoulder dragged.

Since both feet were pinned by moving pins that move along the trajectory determined by the original walking motion, their positions do not change even when we drag other links in the body. We modified the second and third frames by simply dragging the head and left hand so that the motion looks like avoiding an object flying towards the character.

8.6 Summary

This chapter presented a method for generating whole-body motions of human figures by pin-and-drag interface using the newly developed inverse kinematics computation. The contributions of this chapter are summarized as follows:

(1) We developed computational algorithms for the singularity-robust pin-and-drag interface to compute whole-body motions of a human character considering var-

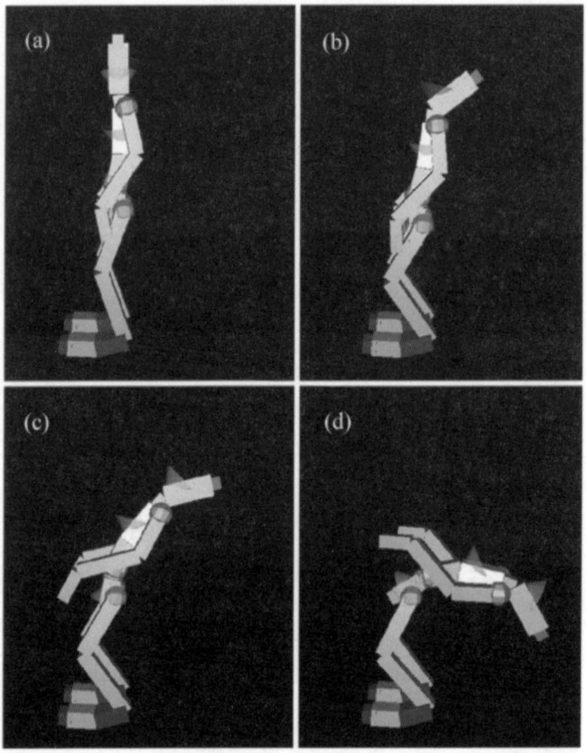

Fig. 8.6. Effect of applying joint motion range; (a) original posture, where the motion ranges are represented by cones, (b) the neck joint exceeds its range, (c) neck joint returns back into the range, (d) the upper body is bended further without exceeding the ranges.

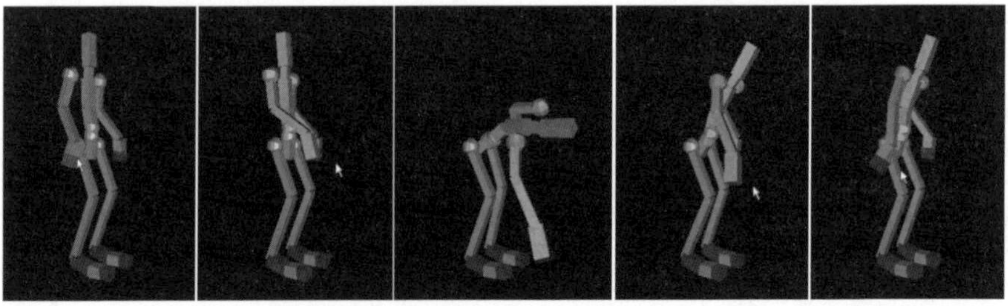

Fig. 8.7. An example of realtime motion generation.

ious kinematic constraints. The resulting motions are quite natural and human-like, probably due to the synergetic effects.

(2) In contrast to the other numerical inverse kinematics solvers, we can place any number of pins to arbitrary links without causing troubles due to the singularity

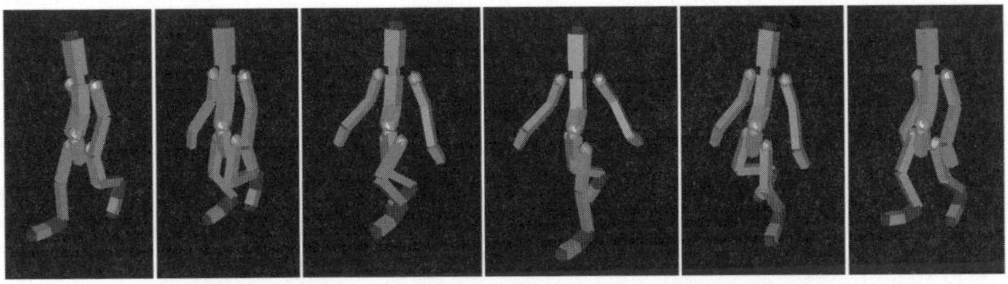

Fig. 8.8. Original motion used for motion editing in motion.

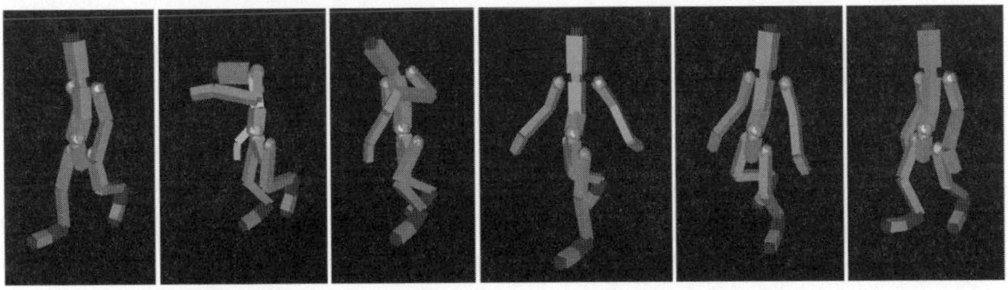

Fig. 8.9. Modified motion from the motion in Fig. 8.8.

of the Jacobian matrix. This feature is enabled by dividing the constraints into two priority levels and applying singularity-robust pseudo inverse.

(3) Implemented constraints include pins, desired joint angles, and joint motion ranges. All of these constraints are maintained by the feedback controllers that compensate the errors due to inconsistencies and/or singularities.

(4) Motion editing in motion and retargetting captured motions are also realized by applying the proposed method with reference motion data.

(5) The computational algorithms were successfully implemented as a software and showed their computational efficiency for realtime choreography without stress of use. Examples of created motions demonstrated the usefulness of the developed algorithms and software.

The method presented in this chapter does not consider any dynamic constraints. The next chapter extends the method to include dynamics to generate physically feasible motions with the same interface.

9

Synergetic Choreography with Dynamics

9.1 Introduction

The interface developed in Chapter 8 considered only the kinematic constraints. As a result, the generated motions tend to appear passive. When the hand is dragged, for example, the figure looks as if it is forced to reach out, rather than it is willing to. In order to generate active motions, we have to compute a whole-body motion that requires no external force at the dragged link. This chapter presents a method to generate active motions with the same interface as in Chapter 8.

The goal here is to find the counterparts for Eqs.(8.9)(8.13) which include the physical constraints as well as the kinematic ones. The joint accelerations are computed using these equations by the same optimization process. Since constraints on joint motion range and desired joint angles are local to each joint and including them would be straightforward, we can focus on the pin and drag constraints. We derive a new equation for pin and drag constraints taking into account of the effects of the contact forces.

The contact forces are computed by the implicit integration technique presented in Chapter 6. The effects of contact forces on the motion of pinned and dragged links are computed by the operational space inertia matrices [15, 48], which are computed by the forward dynamics computation in Chapter 4.

The algorithm is a combination of the new equation of pin and drag constraints and the following three different components that have already been presented in this book:

(1) contact forces computed by the method in Chapter 6, to identify the external forces acting on the figure,
(2) operational inertia matrix computed by the method in Chapter 4, to compute the effect of contact forces on motions of pinned and dragged links, and
(3) feedback controllers in Chapter 8 to maintain the kinematic constraints including joint motion range and desired joint angles.

This chapter is organized as follows. In Section 9.2, it is shown that the operational space inertia matrix is computed by the assembly phase of the forward dynamics computation presented in Chapter 4. Then Section 9.3 derives a new equation for physically consistent pin and drag constraints, which replaces the kinematic constraint used in Chapter 8. The example shown in Section 9.4 demonstrates that the generated motion is dynamically balanced.

9.2 Operational Space Inertia Matrix

Operational space inertia matrix (OSIM) [48] is usually used for the control of manipulators. It describes the relationship between the force applied to an end link and its resulting acceleration, including the effects of the rest of the manipulator. The computation for controlling the trajectory of the end link becomes simple because we can concentrate on the motion and force at the end link. The concept is extended to the extended operational space inertia matrix [15] where relationships between the forces and the motions of m end links are described by m^2 matrices. The extended concept allows simultaneous and/or cooperative control of multiple end links.

While deriving Eq.(4.85), we find the following equations for subchains A and B:

$$ {}^{C}\ddot{\boldsymbol{r}}_{i,A} = {}^{A}\ddot{\boldsymbol{r}}_{i,A} + \boldsymbol{\Lambda}_{Ai,i}(\boldsymbol{K}_{Ci}^{T}\,{}^{C}\boldsymbol{f}_{i} + \boldsymbol{K}_{Ji}^{T}\boldsymbol{\tau}_{i}) \tag{9.1} $$

$$ {}^{C}\ddot{\boldsymbol{r}}_{i,B} = {}^{B}\ddot{\boldsymbol{r}}_{i,B} + \boldsymbol{\Lambda}_{Bi,i}(\boldsymbol{K}_{Ci}^{T}\,{}^{C}\boldsymbol{f}_{i} + \boldsymbol{K}_{Ji}^{T}\boldsymbol{\tau}_{i}) \tag{9.2} $$

which describe the relationship between the acceleration at joint i before and after connecting joint i. Since ${}^{C}\boldsymbol{\tau}_{Gi} \triangleq \boldsymbol{K}_{Ci}^{T}\,{}^{C}\boldsymbol{f}_{i} + \boldsymbol{K}_{Ji}^{T}\boldsymbol{\tau}_{i}$ is the generalized force applied to joint i, $\boldsymbol{\Lambda}_{Ai,i}$ and $\boldsymbol{\Lambda}_{Bi,i}$ are nothing but the OSIM of links p_i and c_i, respectively.

We also derived the following relationship between the acceleration at joint k and the force at joint i:

$$ {}^{C}\ddot{\boldsymbol{r}}_{k,C} = {}^{X}\ddot{\boldsymbol{r}}_{k,X} + \boldsymbol{\Lambda}_{Xk,i}(\boldsymbol{K}_{Ci}^{T}\,{}^{C}\boldsymbol{f}_{i} + \boldsymbol{K}_{Ji}^{T}\boldsymbol{\tau}_{i}) \tag{9.3} $$

where $X \in \{A, B\}$ is selected so that k is also included in $\mathcal{E}_X$, used for computing the acceleration at joint k after assembling joint i. As is obvious from this equation, $\boldsymbol{\Lambda}_{Xk,i}$ is the extended OSIM between joints i and k.

If connections between the environment and several links of the human figures are set, the extended OSIM $\boldsymbol{\Lambda}_E$ is obtained after processing the last joint in the human figure in the assembly phase and satisfies the following equation:

$$ \ddot{\boldsymbol{r}}_E = \boldsymbol{\Lambda}_E \boldsymbol{f}_E + \ddot{\boldsymbol{r}}_E^0 \tag{9.4} $$

where $\boldsymbol{f}_E$ is the force applied to the end links, $\ddot{\boldsymbol{r}}_E^0$ is the temporary acceleration of the end links before assembling the connecting joints, and $\ddot{\boldsymbol{r}}_E$ is the acceleration with the force applied to the end links. Figure 9.1 illustrates the physical meaning of each variable. $\ddot{\boldsymbol{r}}_E^0$ can be interpreted as the acceleration of end links when the figure is flying without interactions with the environment.

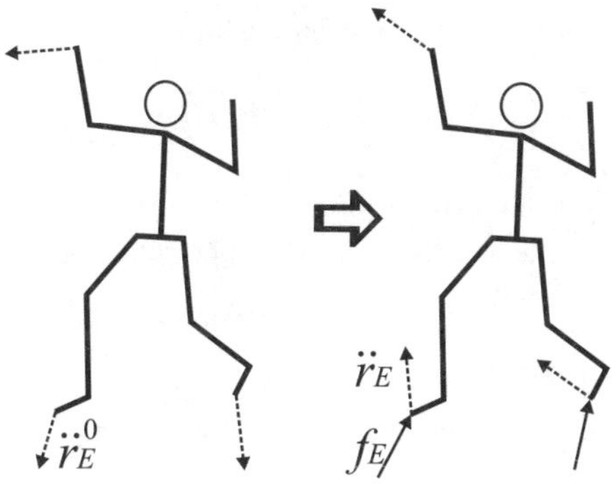

Fig. 9.1. Relationships between the forces and motions of end links.

9.3 Physically Consistent Pin Constraint

Equation 9.4 gives the motions of all end links, provided that the external forces $\boldsymbol{f}_E$ and the accelerations without external forces $\ddot{\boldsymbol{r}}_E^0$ are known. $\ddot{\boldsymbol{r}}_E^0$ depends on the accelerations of all joints. In order to derive the relationship between $\ddot{\boldsymbol{q}}$ and $\ddot{\boldsymbol{r}}_E^0$, we introduce a new matrix ${}^X\boldsymbol{\Phi}_{m,X}$ and a vector ${}^X\boldsymbol{\phi}_{m,X}$ which satisfy the following equation:

$$ {}^X\ddot{\boldsymbol{r}}_{m,X} = {}^X\boldsymbol{\Phi}_{m,X}\ddot{\boldsymbol{q}} + {}^X\boldsymbol{\phi}_{m,X} \tag{9.5} $$

where $\ddot{\boldsymbol{q}}$ is the vector composed of the joint accelerations. ${}^X\boldsymbol{\Phi}_{m,X}$ describes the relationship between the joint acceleration and the acceleration at the end link of subchain X including the effects of forces transferred at all joints in subchain X.

${}^X\boldsymbol{\Phi}_{m,X}$ and ${}^X\boldsymbol{\phi}_{m,X}$ are computed by a recursive procedure. Suppose subchain A is going to be connected to subchain B through joint i to form subchain C and we have already computed ${}^A\boldsymbol{\Phi}_{m,A}$ and ${}^A\boldsymbol{\phi}_{m,A}$ which satisfy

$$ {}^A\ddot{\boldsymbol{r}}_{m,A} = {}^A\boldsymbol{\Phi}_{m,A}\ddot{\boldsymbol{q}} + {}^A\boldsymbol{\phi}_{m,A}. \tag{9.6} $$

Our task is to derive the expression of ${}^C\boldsymbol{\Phi}_{m,C}$ and ${}^C\boldsymbol{\phi}_{m,C}$ such that

$$ {}^C\ddot{\boldsymbol{r}}_{m,C} = {}^C\boldsymbol{\Phi}_{m,C}\ddot{\boldsymbol{q}} + {}^C\boldsymbol{\phi}_{m,C}. \tag{9.7} $$

Accelerations at joint i are also described in the same way as

$$ {}^A\ddot{\boldsymbol{r}}_{i,A} = {}^A\boldsymbol{\Phi}_{i,A}\ddot{\boldsymbol{q}} + {}^A\boldsymbol{\phi}_{i,A} \tag{9.8} $$

$$ {}^B\ddot{\boldsymbol{r}}_{i,B} = {}^B\boldsymbol{\Phi}_{i,B}\ddot{\boldsymbol{q}} + {}^B\boldsymbol{\phi}_{i,B}. \tag{9.9} $$

After adding joint i, the accelerations at joint i changes to

$$ {}^C\ddot{\boldsymbol{r}}_{i,A} = {}^A\ddot{\boldsymbol{r}}_{i,A} + \boldsymbol{\Lambda}_{Ai,i}{}^C\boldsymbol{\tau}_{Gi} \tag{9.10} $$

$$ {}^C\ddot{\boldsymbol{r}}_{i,B} = {}^B\ddot{\boldsymbol{r}}_{i,B} + \boldsymbol{\Lambda}_{Bi,i}{}^C\boldsymbol{\tau}_{Gi}. \tag{9.11} $$

These accelerations should satisfy the relationship

$$\begin{pmatrix} \boldsymbol{K}_{Ji} \\ \boldsymbol{K}_{Ci} \end{pmatrix} \left({}^{C}\ddot{\boldsymbol{r}}_{i,A} + {}^{C}\ddot{\boldsymbol{r}}_{i,B} \right) = \begin{pmatrix} \ddot{\boldsymbol{q}}_i \\ \boldsymbol{O} \end{pmatrix} \tag{9.12}$$

where $\ddot{\boldsymbol{q}}_i$ is the joint acceleration of joint i. ${}^{C}\ddot{\boldsymbol{r}}_{m,C}$ is computed from ${}^{A}\ddot{\boldsymbol{r}}_{m,A}$ using the generalized force ${}^{C}\boldsymbol{\tau}_{Gi}$ and the extended operational space inertia $\boldsymbol{\Lambda}_{Am,i}$ as

$${}^{C}\ddot{\boldsymbol{r}}_{m,C} = {}^{A}\ddot{\boldsymbol{r}}_{m,A} + \boldsymbol{\Lambda}_{Am,i} {}^{C}\boldsymbol{\tau}_{Gi}. \tag{9.13}$$

Solving Eqs.(9.10)–(9.12) yields

$${}^{C}\boldsymbol{\tau}_{Gi} = \boldsymbol{P}_i^{-1} (\bar{\boldsymbol{K}}_{Ji} \ddot{\boldsymbol{q}}_i - {}^{A}\ddot{\boldsymbol{r}}_{i,A} - {}^{B}\ddot{\boldsymbol{r}}_{i,B}) \tag{9.14}$$

where

$$\left(\bar{\boldsymbol{K}}_{Ji} \ \bar{\boldsymbol{K}}_{Ci} \right) = \begin{pmatrix} \boldsymbol{K}_{Ji} \\ \boldsymbol{K}_{Ci} \end{pmatrix}^{-1} \tag{9.15}$$

$$\boldsymbol{P}_i = \boldsymbol{\Lambda}_{Ai,i} + \boldsymbol{\Lambda}_{Bi,i}. \tag{9.16}$$

Comparing Eq.(9.7) with the result obtained by substituting Eq.(9.14) into Eq.(9.13) and using Eqs.(9.6)(9.8)(9.9) yields

$${}^{C}\boldsymbol{\Phi}_{m,C} = {}^{A}\boldsymbol{\Phi}_{m,A} + \boldsymbol{\Lambda}_{Am,i} \boldsymbol{P}_i^{-1} (\boldsymbol{\Phi}_{Ji} - {}^{A}\boldsymbol{\Phi}_{i,A} - {}^{B}\boldsymbol{\Phi}_{i,B}) \tag{9.17}$$

$${}^{C}\boldsymbol{\phi}_{m,C} = {}^{A}\boldsymbol{\phi}_{m,A} - \boldsymbol{\Lambda}_{Am,i} \boldsymbol{P}_i^{-1} ({}^{A}\boldsymbol{\phi}_{i,A} + {}^{B}\boldsymbol{\phi}_{i,B}) \tag{9.18}$$

where

$$\boldsymbol{\Phi}_{Ji} \stackrel{\triangle}{=} \left(\boldsymbol{O} \ldots \bar{\boldsymbol{K}}_{Ji} \ldots \boldsymbol{O}. \right). \tag{9.19}$$

Repeating this procedure until the human figure is completed, we obtain $\boldsymbol{\Phi}_E$ and $\boldsymbol{\phi}_E$ which satisfy

$$\ddot{\boldsymbol{r}}_E^0 = \boldsymbol{\Phi}_E \ddot{\boldsymbol{q}} + \boldsymbol{\phi}_E. \tag{9.20}$$

Substituting this equation to Eq.(9.4) we have

$$\ddot{\boldsymbol{r}}_E = \boldsymbol{\Lambda}_E \boldsymbol{f}_E + \boldsymbol{\Phi}_E \ddot{\boldsymbol{q}} + \boldsymbol{\phi}_E. \tag{9.21}$$

Suppose a pin is set at link i, whose acceleration is denoted by $\ddot{\boldsymbol{r}}_{Ei}$, and link k is in contact with the environment. Then, the following equation is derived from Eq.(9.21):

$$\ddot{\boldsymbol{r}}_{Ei} = \boldsymbol{\Lambda}_{Ei,k} \boldsymbol{f}_{Ek} + \boldsymbol{\Phi}_{Ei} \ddot{\boldsymbol{q}} + \boldsymbol{\phi}_{Ei} \tag{9.22}$$

where $\boldsymbol{f}_{Ek}$ is the contact force/moment computed in the way described in Chapter 6, and $\boldsymbol{\Phi}_{Ei}$ and $\boldsymbol{\phi}_{Ei}$ are the rows of $\boldsymbol{\Phi}_E$ and $\boldsymbol{\phi}_{Ei}$ corresponding to link i, respectively.

Given the desired acceleration $\ddot{\boldsymbol{r}}_{Ei}^d$ of the pinned link i, the joint accelerations should satisfy

$$\boldsymbol{\Phi}_{Ei} \ddot{\boldsymbol{q}} = \ddot{\boldsymbol{r}}_{Ei} - \boldsymbol{\Lambda}_{Ei,k} \boldsymbol{f}_{Ek} - \boldsymbol{\phi}_{Ei}. \tag{9.23}$$

Equation (9.23) is the physically consistent version of pin constraint that replaces Eq.(8.9). If the pin constraint is given the lower priority, this equation is to be included in Eq.(8.13). Other constraints, joint motion range and desired joint angles, are handled in a similar way as in Chapter 8.

$\ddot{\boldsymbol{q}}$ is computed by the same optimization process as in Chapter 8, which is then integrated to compute the joint motion.

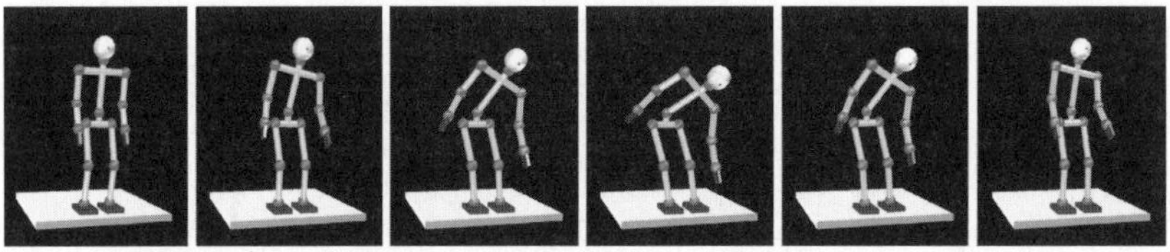

Fig. 9.2. A motion generated by pin-and-drag interface with dynamics.

9.4 Example

Fig. 9.2 shows a motion generated by pin-and-drag interface with dynamics. The left hand was pulled down for 2.5 s while the both feet were in contact with the ground. There were also weak pins at both feet that constrain the orientations. Notice that the pelvis moves to right in order to maintain balanced. In the motion of Fig. 8.7, in contrast, the pelvis moves little.

9.5 Summary

This chapter presented an online motion generation technique that combines the pin-and-drag interface of Chapter 8 and the efficient dynamics computation of Chapters 4 and 6. The method is capable of generating physically consistent whole-body motions with the intuitive pin-and-drag interface.

The relevant quantities are the operational space inertia matrix $\boldsymbol{\Lambda}_E$ and $\boldsymbol{\Phi}_E$ that relates the accelerations of the joints and end links with no physical interactions with the environment. These quantities enable us to ignore the assembly procedure to build the human figure and concentrate on the motions of end links. This property naturally fits the pin-and-drag interface where only the motions of selected end links are in question, rather than the motions of individual joints.

The problem of the method comes from the lack of controller as discussed in Chapter 7. Once a figure fails to keep balanced, it never recovers a stable posture because it simply tries to return to the fixed desired positions or angles. In order to adapt the method to a wide range of situation, we would need a powerful controller that controls the desired values based on a strategy to recover global stability.

10

Controlling a Marionette with Human Motion Capture Data

10.1 Introduction

Entertainment is one of the more immediately practical applications of humanoid robots and several robots have recently been developed for this purpose [35, 45, 90]. In this chapter, we explore the use of an inexpensive entertainment robot controlled by motion capture data with the goal of making such robots readily available and easily programmable. The robot is a marionette where the length and pivot points of the strings are controlled by eight servo motors that bring the hands and the feet of the marionette to the desired positions (Fig. 10.1).

Standard marionettes are puppets with strings operated by a human performer's hands and fingers. Creating appealing motion with such a puppet is difficult and requires extensive practice. Although for our marionette the servo motors move the strings, programming a robotic version of such a device by hand to produce expressive gestures would also be difficult. We solve this problem by using full-body human motion data to drive the motion. The human motion data is recorded as the positions of markers in three dimensions while an actor tells a story with his or her hands. After adaptation, the data are used to drive the motion of the marionette by taking into account the marionette's swing dynamics. These adaptations and dynamics compensation are necessary because the marionette has many fewer degrees of freedom and much smaller size than the human actor, strings rather than actuators at the joints, and no sensors except for the rotation of the motors.

The method described in this chapter consists of four steps: (1) identify the swing dynamics of the hands and design a feedforward controller to prevent swinging and obtain a desired response, (2) obtain the translation, orientation, and scaling parameters that map the measured marker positions for the human motion into the marionette's workspace, (3) apply the controller to modify the mapped marker positions

---

This chapter was adapted from, by permission, K. Yamane, J.K. Hodgins, and H.B. Brown, "Controlling a Marionette with Human Motion Capture Data," Proceedings of IEEE International Conference on Robotics and Automation, Taipei, Taiwan, September 2003 (in print).

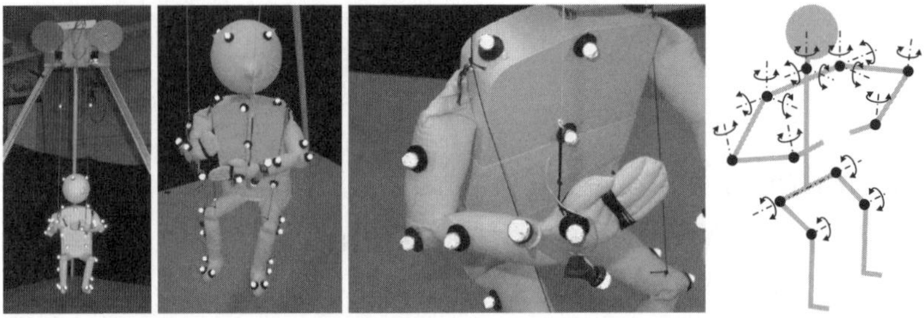

Fig. 10.1. The motor-driven marionette and its model. The marionette is about 60cm tall. The shoulder and elbow joints have cloth stops to prevent unrealistic joint angles.

to prevent swing, and (4) compute the motor commands to bring the (virtual) markers attached to the marionette to the revised positions computed in step (3).

The relationship between the four steps is illustrated in Fig. 10.2. In steps (2) and (4), we solve the inverse kinematics problem with many different constraints including marker positions, joint motion ranges, strings, and gravity. This algorithm is an extension of the method in Chapter 8. In step (1), we model the dynamics of swing by capturing the response to a step input of each desired marker position and then use that response to design a feedforward controller to compensate for the swing motion.

As a demonstration of the algorithm, we include experimental results to compare the marionette motion to that of human actors. The motions of the marionette and the human actor are similar enough to distinguish different styles for the same story.

10.2 Related Work

Hoffmann [33] developed a human-scale marionette and controlled it to perform dancing motions using human data. The size and controllable degrees of freedom of the marionette are much closer to those of human than ours. The research is therefore focused more on image processing for measuring human motion than on mapping between human and marionette motions.

Mapping motion data is a common problem in applying motion capture data to a real robot or to a virtual character. The factors considered in previous work include joint angle and velocity limits [82], kinematic constraints [28], and physical consistency [94]. However, the original and target human figures are usually more similar in degrees of freedom, dimensions, and actuators than the marionette is to a human actor.

The mechanism and dynamics of a string-driven marionette are quite similar to those of wired structures such as a crane. A number of researchers have worked on controlling a crane to bring an object to a desired position without significant oscillations [85]. This work assumes that the position of the object is known through the direction of the wire. Although we measure the position of the hand and feet

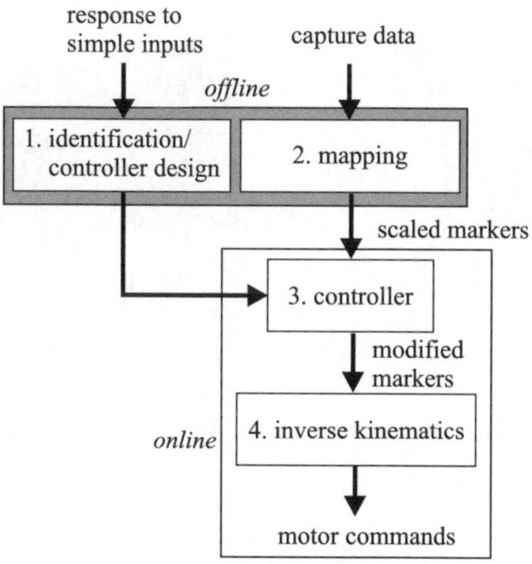

Fig. 10.2. Overview of the marionette control system. The blocks in the top gray box are processed offline: identification/controller design for each marionette and mapping for each motion sequence. Those in the bottom white box is processed in realtime during a performance.

for recording the swing dynamics, we do not have this information during a performance. Our accuracy requirements are much less because the marionette is gesturing in free space rather than precisely positioning an object.

10.3 Experimental Setup

The marionette is modeled as a 17DOF kinematic chain (Fig. 10.1 right). Closeups of the motors and pulleys are shown in Fig. 10.3. The marionette has eight servo motors (Airtronics Servo 94102); six control the arms and two control the legs. The motors are controlled by position commands sent from a serial port of a PC via an 8-channel controller board (Pontech SV203).

Motors 3 and 6 change the length of the strings connecting the hands. Motors 7 and 8 move the knee up and down. Motors 1, 2, 4 and 5 move the hands in horizontal directions by rotating the "pipes" and moving the pipe ends via four independent planar linkages (Fig. 10.4).

We used a commercially available motion capture system from Vicon for capturing the actor's performance and the marionette's motion for identification of the swing dynamics. The system has nine cameras, each capable of recording images of 1000×1000 pixels resolution at 120Hz. We used different camera placements for the human subject and the marionette to accommodate the smaller workspace of the marionette and to ensure accurate measurements.

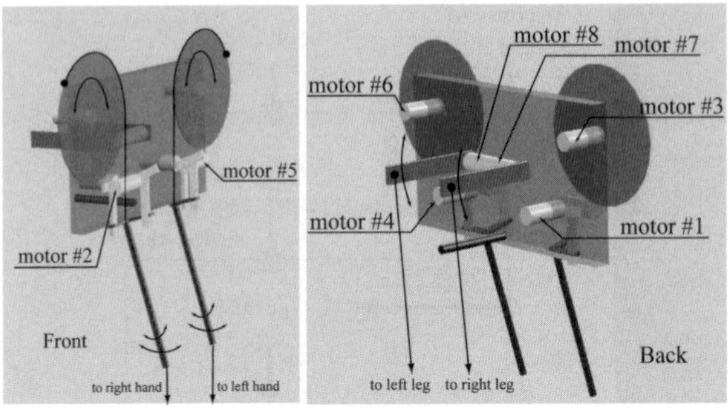

Fig. 10.3. Closeup of the motors and pulleys; front of marionette (left), back (right).

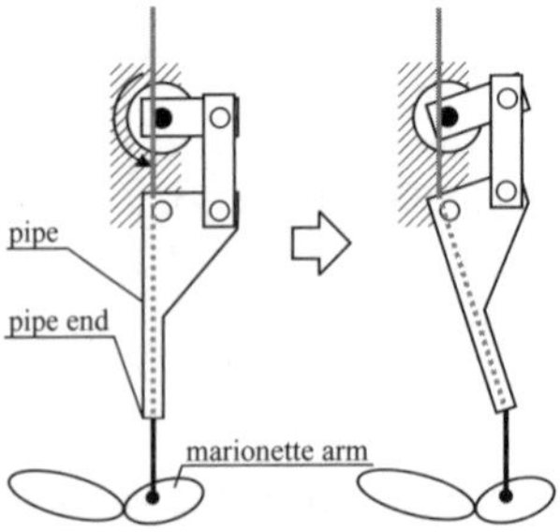

Fig. 10.4. The mechanism for moving the hand in the horizontal direction.

10.4 Inverse Kinematics

The inverse kinematics algorithm is used to enforce constraints to bring the markers representing the desired motion into the workspace of the marionette and to determine the motor angles that satisfy the desired marker positions and the physical constraints, including the desired marker positions, joint motion ranges, length and orientation of the strings, and potential energy. The potential energy constraint is introduced to model the effect of gravity. The inverse kinematics algorithm computes the joint angles and the motor commands that locally optimize the constraints. Because the algorithm was described in Chapter 8, we present a short outline here.

Often, all the constraints cannot be satisfied due to the singularity of the configuration or to inconsistencies between the constraints. Therefore, the user is asked to divide the constraints into two groups: those that must be satisfied and those where some error is acceptable. The algorithm applies singularity-robust (SR) inverse [64] (also known as damped pseudo inverse [54]) to the lower-priority constraints. As described below, the SR-inverse distributes the error among the lower-priority constraints according to the given weights so that the resulting joint velocity does not become too large even if there are singularities or inconsistencies in the constraints.

We design a feedback controller for each constraint to ensure that the lower-priority constraints are satisfied as much as possible and to eliminate integration errors in both higher- and lower-priority constraints. The controller computes the required velocity when constraints are violated. For example, the feedback controller to bring a link to its reference position $\boldsymbol{p}^{ref}$ is $\dot{\boldsymbol{p}}^{des} = k_p(\boldsymbol{p}^{ref} - \boldsymbol{p})$ where k_p is a positive gain, $\boldsymbol{p}$ is the current position, and $\dot{\boldsymbol{p}}^{des}$ is the desired velocity. Note that this velocity is not always realized for lower-priority constraints due to the nature of the SR-inverse algorithm.

With n_1 higher-priority constraints and n_2 lower-priority constraints, we have the following equations in generalized velocity $\dot{\boldsymbol{\theta}}$:

$$\boldsymbol{J}_1\dot{\boldsymbol{\theta}} = \boldsymbol{v}_1^{des} \tag{10.1}$$

$$\boldsymbol{J}_2\dot{\boldsymbol{\theta}} = \boldsymbol{v}_2^{des} \tag{10.2}$$

where $\boldsymbol{v}_1^{des}$ and $\boldsymbol{v}_2^{des}$ are the desired velocities corresponding to higher- and lower-priority constraints respectively, and $\boldsymbol{J}_1$ and $\boldsymbol{J}_2$ are the Jacobian matrices of the constraints with respect to $\boldsymbol{\theta}$.

We solve this equation for the generalized velocity as follows. First, we compute the set of exact solutions of Eq.(10.1) by

$$\dot{\boldsymbol{\theta}} = \boldsymbol{J}_1^{\sharp}\boldsymbol{v}_1^{des} + (\boldsymbol{I} - \boldsymbol{J}_1^{\sharp}\boldsymbol{J}_1)\boldsymbol{y} \tag{10.3}$$

where $\boldsymbol{J}_1^{\sharp}$ is the pseudoinverse of $\boldsymbol{J}_1$, $\boldsymbol{I}$ is the identity matrix of the appropriate size, and $\boldsymbol{y}$ is an arbitrary vector. We can rewrite this equation as $\dot{\boldsymbol{\theta}} = \dot{\boldsymbol{\theta}}_1 + \boldsymbol{W}\boldsymbol{y}$ where $\dot{\boldsymbol{\theta}}_1 \stackrel{\triangle}{=} \boldsymbol{J}_1^{\sharp}\boldsymbol{v}_1^{des}$, and $\boldsymbol{W} \stackrel{\triangle}{=} \boldsymbol{I} - \boldsymbol{J}_1^{\sharp}\boldsymbol{J}_1$. Next, we compute the $\boldsymbol{y}$ with which $\dot{\boldsymbol{\theta}}$ would satisfy Eq.(10.2) as closely as possible by

$$\boldsymbol{y} = (\boldsymbol{J}_2\boldsymbol{W})^*(\boldsymbol{v}_2^{des} - \boldsymbol{J}_2\dot{\boldsymbol{\theta}}_1) \tag{10.4}$$

where $(\boldsymbol{J}_2\boldsymbol{W})^*$ is the SR-inverse of $\boldsymbol{J}_2\boldsymbol{W}$. Finally, the generalized velocity $\dot{\boldsymbol{\theta}}$ is computed by substituting $\boldsymbol{y}$ into Eq.(10.3), which is then integrated to compute the generalized coordinates in the next step.

In order to add a new constraint, we must design a feedback controller to compute the desired velocity and derive the corresponding Jacobian matrix. In the following subsections, we describe the string and potential energy constraints in detail. We used the equations in Chapter 8 for other constraints.

10.4.1 String Constraints

Each string has a start point, an end point, and some number of intermediate points (Fig. 10.5 left). The string can slide back and forth at the intermediate points. The current length of a string, l, must always be equal to or smaller than its nominal length l_0. l is computed by summing the length of all segments:

$$l = \sum_{i=0}^{N-1} l_i = \sum_{i=0}^{N-1} |\boldsymbol{p}_{i+1} - \boldsymbol{p}_i|$$

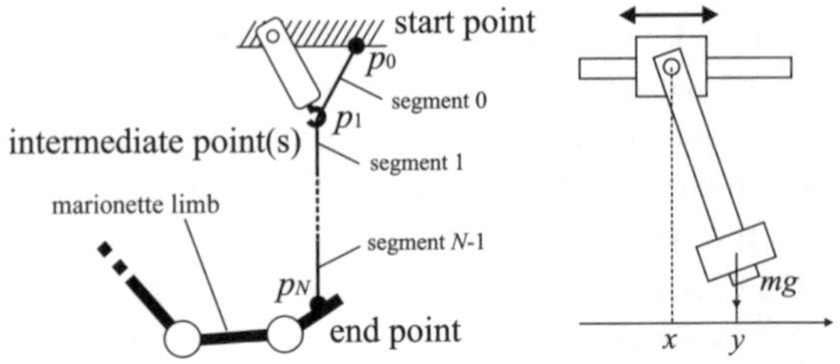

Fig. 10.5. String models for inverse kinematics (left) and swing controller (right).

$$= \sum_{i=0}^{N-1} \sqrt{(\boldsymbol{p}_{i+1} - \boldsymbol{p}_i)^T (\boldsymbol{p}_{i+1} - \boldsymbol{p}_i)} \tag{10.5}$$

where N is the number of segments, l_i $(0 \leq i \leq N-1)$ is the length of segment i, $\boldsymbol{p}_i$ $(0 \leq i \leq N)$ is the position of the i-th point. The Jacobian matrix of l with respect to the generalized coordinates $\boldsymbol{\theta}$ is computed by

$$\begin{aligned} \boldsymbol{J}_{str} &= \frac{\partial l}{\partial \boldsymbol{\theta}} = \sum \frac{\partial l_i}{\partial \boldsymbol{\theta}} \\ &= \sum \frac{1}{l_i} (\boldsymbol{p}_{i+1} - \boldsymbol{p}_i)^T \left(\frac{\partial \boldsymbol{p}_{i+1}}{\partial \boldsymbol{\theta}} - \frac{\partial \boldsymbol{p}_i}{\partial \boldsymbol{\theta}} \right) \\ &= \sum \frac{1}{l_i} (\boldsymbol{p}_{i+1} - \boldsymbol{p}_i)^T (\boldsymbol{J}_{i+1} - \boldsymbol{J}_i). \end{aligned} \tag{10.6}$$

Note that the Jacobian matrix is not defined for segments with $l_i = 0$, although we never encounter such situations in physical mechanisms. The feedback law for a string constraint is $v_{str}^{des} = k_{str}(l_0 - l)$.

In addition to the length, we also constrain the $(N-1)$th segment of the string to be vertical due to gravity. The two points $\boldsymbol{p}_{N-1}$ and $\boldsymbol{p}_N$ should then be vertical:

$$\boldsymbol{h}_1 \cdot \boldsymbol{d}_{N-1,N} = 0, \ \boldsymbol{h}_2 \cdot \boldsymbol{d}_{N-1,N} = 0 \tag{10.7}$$

where $\boldsymbol{h}_1$ and $\boldsymbol{h}_2$ are independent unit vectors in the horizontal plane (e.g. $\boldsymbol{h}_1 = (1\ 0\ 0)^T, \boldsymbol{h}_2 = (0\ 1\ 0\ )^T$ if the gravity is in z direction) and $\boldsymbol{d}_{N-1,N}$ is the unit vector from $\boldsymbol{p}_{N-1}$ to $\boldsymbol{p}_N$, namely $\boldsymbol{d}_{N-1,N} = (\boldsymbol{p}_N - \boldsymbol{p}_{N-1})/l_{N-1}$. The Jacobian matrix for this constraint is

$$\boldsymbol{J}_v = \frac{1}{l_{N-1}} \begin{pmatrix} \boldsymbol{h}_1^T \\ \boldsymbol{h}_2^T \end{pmatrix} (\boldsymbol{J}_N - \boldsymbol{J}_{N-1}). \tag{10.8}$$

The desired velocity for this constraint is

$$\boldsymbol{v}_v^{des} = -k_v \begin{pmatrix} \boldsymbol{h}_1^T \\ \boldsymbol{h}_2^T \end{pmatrix} \boldsymbol{d}_{N-1,N} \tag{10.9}$$

where k_v is a positive gain.

10.4.2 Potential Energy

Because the joints that do not have strings directly attached to them will bend downward due to gravity, we also constrain the potential energy to be as small as possible by constraining the center of mass of the whole body to be as low as possible. The Jacobian matrix for this constraint is computed as $\boldsymbol{J}_{pe} = \boldsymbol{d}_g^T \boldsymbol{J}_{COM}$ where $\boldsymbol{d}_g = (0\ 0\ 1)^T$ is the unit vector in the direction of the gravity and $\boldsymbol{J}_{COM}$ is the Jacobian matrix of the center of mass with respect to the generalized coordinates. A method for computing $\boldsymbol{J}_{COM}$ can be found in [92]. The desired velocity for this constraint is a negative constant $-k_{pe}$.

10.5 Mapping

Before applying the measured marker positions to a marionette, we need to map them into new positions not only to adapt to the size of the marionette but also to comply with such physical constraints as the strings. Our marionette, for example, does not have a mechanism to move the pelvis. Therefore, if the captured motion contains translation or rotation of the pelvis, the motion should be translated or rotated so that the pelvis motion is eliminated.

In this section, we describe an algorithm to compute seven parameters for translation, rotation, and scaling that map the measured marker positions into new positions that satisfy the constraints of the inverse kinematics model described in Section 10.4. We compute the mapping parameters independently for each frame rather than using the same parameters for all frames. Although it might seem natural to fix a parameter such as scaling for a particular human actor, we have found that, because of the marionette's limited range of motion, using the best possible mapping for each frame is preferable to using a fixed mapping bounded by the most difficult posture in the motion clip.

Suppose we use N markers in frame k as reference and denote the positions of the markers attached to the marionette by $\boldsymbol{p}_{k,i}^M$, those of the captured markers by $\boldsymbol{p}_{k,i}^C$ and those of the mapped markers by $\boldsymbol{p}_{k,i}^S (i = 1 \dots N)$. We represent the translational, rotational, and scaling parameters of frame k by a position vector $\boldsymbol{t}_k$, a 3-by-3 rotation matrix $\boldsymbol{R}_k$, and a scalar s_k, respectively. Using these parameters, we compute the mapped position $\boldsymbol{p}_{k,i}^S$ of marker i from its original captured position $\boldsymbol{p}_{k,i}^C$ as $\boldsymbol{p}_{k,i}^S = s_k \boldsymbol{R}_k \boldsymbol{p}_{k,i}^C + \boldsymbol{t}_k$.

The system first computes the scaling, translation, and orientation parameters that minimize the total square distance between the measured markers and the virtual markers on the marionette in a fixed configuration. We can then use the inverse kinematics algorithm to compute the joint angles and string lengths that provide the best match between the two sets of markers. These two steps are repeated a number of times to refine the result.

The system finds the translation, rotation, and scaling parameters $\boldsymbol{t}_k$, $\boldsymbol{R}_k$, and s_k that minimize the evaluation function

$$J_k = \frac{1}{2}\sum_{i=1}^{N} |\boldsymbol{p}^S_{k,i} - \boldsymbol{p}^M_{k,i}|^2. \tag{10.10}$$

$\boldsymbol{p}^M_{k,i}$ are constant because the configuration of the marionette is fixed during this frame by frame computation.

We combine the unknowns into one variable $\boldsymbol{q}_k \in \boldsymbol{R}^7$, where the rotation matrix is represented by three independent variables whose time derivatives correspond to the angular velocity. We then use the common gradient method [84] to compute the optimum $\boldsymbol{q}_k$ incrementally as

$$\boldsymbol{q}_k = \boldsymbol{q}_k + \Delta\boldsymbol{q}_k,\ \Delta\boldsymbol{q}_k = -k\frac{\partial J_k}{\partial \boldsymbol{q}_k}. \tag{10.11}$$

The partial derivative of the mapped position $\boldsymbol{p}^S_{k,i}$ with respect to $\boldsymbol{q}_k$ is computed as

$$\boldsymbol{H}_{i,k} \triangleq \frac{\partial \boldsymbol{p}^S_{k,i}}{\partial \boldsymbol{q}_k} = \left(\boldsymbol{I}_3 \middle| [\boldsymbol{r}\times] \middle| \boldsymbol{R}_k \boldsymbol{p}^C_{k,i}\right) \tag{10.12}$$

where $\boldsymbol{I}_3$ is the 3-by-3 identity matrix, $\boldsymbol{r} \triangleq s_k \boldsymbol{R}_k \boldsymbol{p}^C_{k,i}$, and $[\boldsymbol{r}\times]$ is the cross product matrix of $\boldsymbol{r}$. Using $\boldsymbol{H}_{k,i}$, the partial derivative of J_k is computed by

$$\frac{\partial J_k}{\partial \boldsymbol{q}_k} = (\boldsymbol{p}^S_{k,i} - \boldsymbol{p}^M_{k,i})^T \boldsymbol{H}_{k,i}. \tag{10.13}$$

We apply this process to each frame independently by starting from the same initial guess. Using the result of previous frame as the initial guess would reduce the computation time, but the algorithm might not recover from a failure to obtain good mapping parameters in one frame due to, for example, missing markers. Regardless of the initial guess, the resulting mapping parameters may not be continuous because the algorithm is finding only a local minima. Small discontinuities are not a problem, however, because the marker positions are "filtered" by the feedback controller and by the SR-inverse used in the inverse kinematics computation.

10.6 Controlling Swing

If the mapped motion is applied directly to the marionette, the hands of the marionette will swing and the motion will not be a good match to that of the human actor. We solve this problem by building a simple linear model for the swing dynamics and experimentally identifying its parameters. An alternative approach would be to model the full dynamics of the marionette, but this tactic is not practical because of uncertainty in the model parameters and the limitations of the motors and sensors. Because marionette is made of wood and cloth, it is difficult to precisely determine the mass, inertia, and friction parameters of the joints. The joints are cleverly designed to prevent unrealistic joint angles (Fig. 10.1), but this design also makes modeling of the system more difficult. The motors are inexpensive hobby servos and do not provide precise control. Furthermore, we do not have sensors that measure the current state of the marionette during a performance.

For the simple model of the dynamics, we make three assumptions. (1) Swinging of the hands occurs in the horizontal plane. Pulling the hands or legs up or down does not create a swinging motion. (2) The motion of a hand along the x axis (left/right) and the y axis (forward/back) are independently controlled by one motor each. (3) There is no coupling between the swinging of the left and right hands. These simplifying assumptions allow us to model swing as four independent systems, two for each hand.

Some situations occur in which the second and third assumptions do not hold. The hand marker sometimes moves along a circular trajectory rather than a straight line. The markers with fixed inputs inevitably move slightly when other markers are moved, violating the last assumption. Both problems are most likely to occur when the hand is relatively close to the body because the stiffness of the elbow and shoulder joints forces the hand away from the body.

10.6.1 Modeling of Swing Dynamics

In this section we describe the swing dynamics model that, when combined with the feedback controller of the inverse kinematics algorithm in Section 10.4, predicts the swing motion.

The inverse kinematics algorithm included a proportional controller, where the velocity of the pipe end $\dot{x}$ is computed from the current position of the pipe end x and the desired position u as $\dot{x} = k(u - x)$ where k is a constant gain. Therefore the transfer function from the marker trajectory to the motion of the pipe end takes the form

$$x = \frac{1}{a_{ik}s + 1}u \tag{10.14}$$

where u is the input (marker trajectory), x is the output (motion of the pipe end), s is the Laplace transformation operator, and a_{ik} is the parameter that determines the amount of delay.

The motion of the hand for a given trajectory of the pipe end can be modeled as a pendulum with a moving base (Fig. 10.5 right). Using the length of the pendulum l and the damping term d, the equation of motion of the pendulum under gravity g is linearized around $x = y$ as $\ddot{y} = l/g(x - y) + d(\dot{x} - \dot{y})$. In general, therefore, the transfer function from the motion of the pipe end to the marker motion is written as

$$y = \frac{b_s s + 1}{a_s s^2 + b_s s + 1}x \tag{10.15}$$

where y is the output (actual marker trajectory) and a_s and b_s are the parameters that determine the frequency and damping respectively.

Combining Eqs.(10.14) and (10.15), the system computing the desired marker trajectory from the actual trajectory will be a 3rd-order system. We estimate the three parameters from motion capture data.

The gains of both systems were assumed to be 1, which turned out to be not true, probably because the joint motion ranges of the pipe prevented the pipe end from reaching the desired position or because the stiffness of the arm joints did not allow the string to be perpendicular. We decided not to consider these model errors because the desired marker position is not achievable if it violates the joint range constraint and the stiffness strongly depends on the configuration of the arm making the system too complicated.

10.6.2 Feedforward Controller

The feedforward controller K is formed by connecting the desired response G_D and the inverse of the estimated model P_m in series, that is, $K = G_D P_m^{-1}$. In order for the controller to be proper (the order of the denominator of the transfer function is larger than that of the numerator), the order of G_D must be larger than 2. We selected a 3rd-order G_D so that the output of the controller is continuous. We can also improve the response of the total system by selecting G_D with a smaller delay. In practice, however, we cannot use an arbitrarily fast G_D because as the gain of the controller increases, it becomes sensitive to modeling errors.

The parameters of the string dynamics model, a_s and b_s, depend on the length of the strings; therefore, we repeat the identification process for several different heights for each hand and design a controller for each model. We then apply the weighted sum of the outputs of the three controllers, where the weights are determined according to the actual height during a performance.

10.7 Results

The inverse kinematics computation to obtain the motor commands was repeated four times for each frame to ensure convergence. The total computation time was about 36 ms per frame on a laptop PC with a Mobile PentiumIII 1GHz processor. Motor commands were sent every 50 ms.

Based on the inverse kinematics computation, we developed an online control interface for the marionette. The model consists of nine string length constraints, joint motion range constraints for eight joints, two string direction constraints, and the potential energy constraint. The user can select a marker and drag it to any position. The inverse kinematics algorithm then computes the motor commands and the joint angles to move the marker to the specified position. Fig. 10.6 shows several snapshots of the marionette model and the corresponding postures of the actual marionette.

The swing controller was designed for three different heights (-0.59m, -0.44m, and -0.29m, measured from the center of the panel where the motors and pulleys are attached). We had a total of twelve controllers for the x and y directions of both hands. Table 10.1 lists the parameter sets for the right hand in the x direction. The parameters were tuned manually, although it should also be possible to apply standard system identification techniques [49].

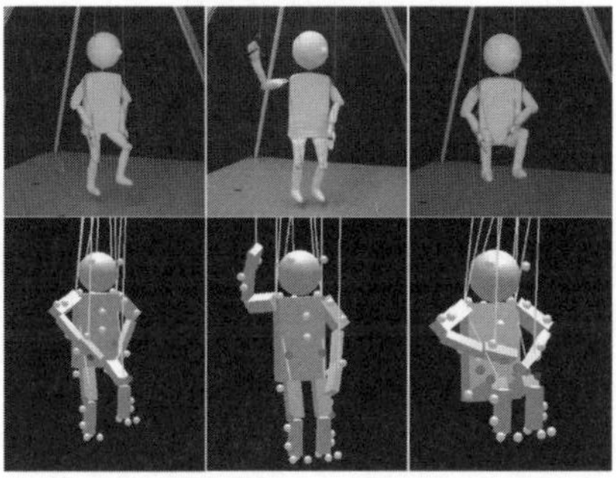

Fig. 10.6. Postures generated by the interactive interface. Above: marionette, below: simulation.

Table 10.1. Parameters of the string dynamics models for the x direction of the right hand.

height [m]	-0.59	-0.44	-0.29
a_{ik}	0.8		
a_s	0.063	0.061	0.059
b_s	0.07	0.06	0.02

Figs. 10.7 and 10.8 show the results of identification, controller design, and verification processes at the height of -0.29m. We used the motion capture system to measure the motion of the pipe end and right hand when a step input in x direction (left to right) was given as the desired marker trajectory (Fig. 10.7). Then we designed a swing controller with the desired response $G_D = 1/(0.2s + 1)^3$.

Finally, the designed controller was applied to the same desired marker trajectory used for the identification and the response was measured (Fig. 10.8). The swing controller reduced the width of the first vibration by 40%. The trajectory of the hand without the swing controller is different from that used for parameter identification (Fig. 10.7), although we used the same reference trajectory. This discrepancy probably explains why the controller could not remove the vibration completely, thereby illustrating that a small difference in the configuration results in a relatively large difference in the swing dynamics due to the stiffness of the arms.

To test the motion of the marionette on a longer performance, we recorded the motions of two actors for two stories: "Who Killed Cockrobin?" and "Alaska." Fig. 10.9 compares the motions based on "Alaska" performed by actor 1. We used 32 reference markers and the two steps for mapping (computing approximate parameters and computing exact parameters) were repeated up to 500 times at each frame. The iteration was suspended if the total error of the marker positions were larger than the previous iteration. The computation time was approximately 5 seconds per frame.

Figs. 10.9 and 10.10 illustrate the same story performed by two different actors. The gestures are taken from approximately the same point in the story. The motion in Fig. 10.11 is based on a different story performed by actor 1. The video

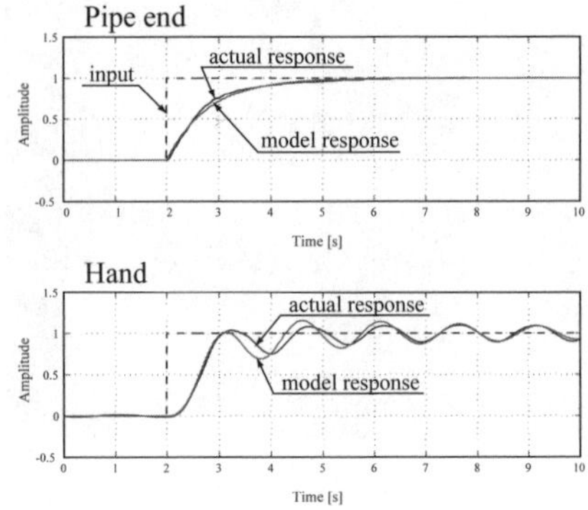

Fig. 10.7. Actual and model responses to a step input. The amplitude of each motion is normalized. The hand of the marionette comes close to its head at this height.

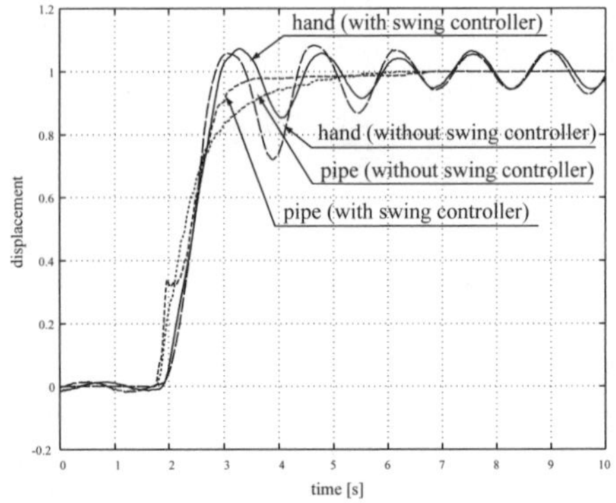

Fig. 10.8. Response of the pipe and the hand to a step input.

clips are available online at http://humanoids.cs.cmu.edu/projects/marionette/.html, which also includes comparisons between the motions with and without the swing controller. The marionette's feet touch the floor as in a real performance.

10.8 Discussion

The motions of the actor and the marionette showed good correspondence, and we were able to distinguish two different styles for the same story (Figs. 10.9 and 10.10). However, significant differences between the actor's and the marionette's postures were sometimes visible because of the limited range of motion of the pipes (for example the middle column of Fig. 10.9). The marionette also had difficulty with fast motions because of the latency in the feedback controller of the inverse kinematics computation. This problem could be solved with a faster computer that could execute

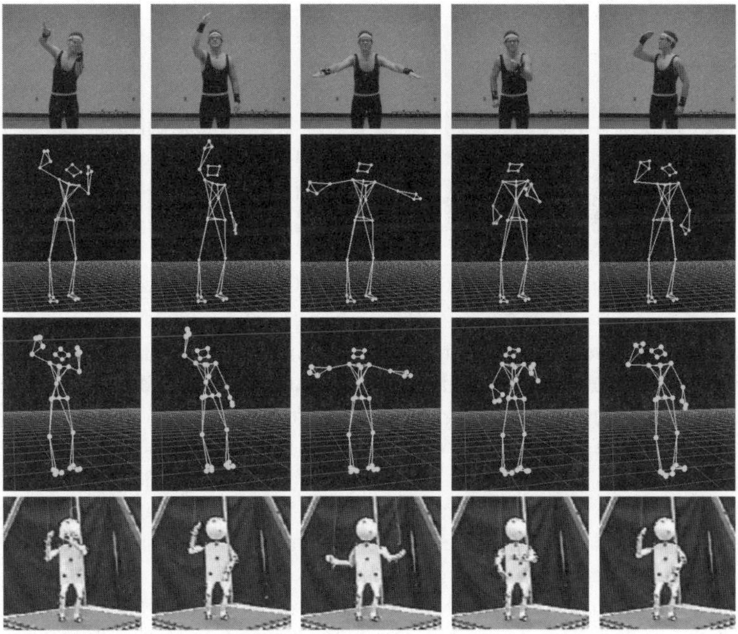

Fig. 10.9. From the top: performance of actor 1 for "Alaska," the motion capture data, mapped marker positions, and the marionette's motion.

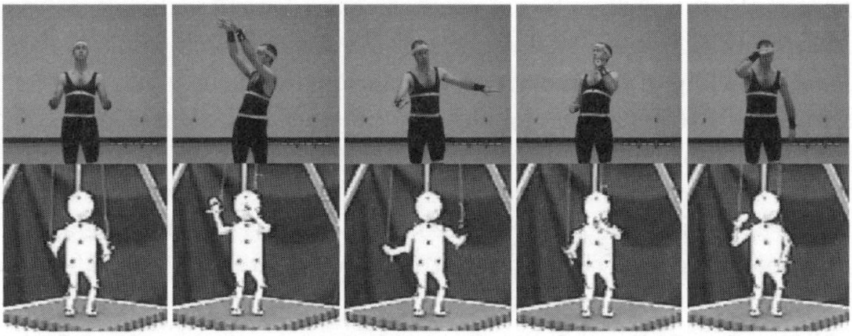

Fig. 10.10. Marionette's motion for "Alaska" performed by actor 2.

more iterations per step of the inverse kinematics computation, thereby increasing the stability of the computation and allowing larger gains.

Although the swing controller had a significant effect in isolated experiments, its effect during longer performances was quite small. We believe this discrepancy occurred because the stiffness of the arms is highly dependent on the configuration and this effect was not taken into account in the swing model. We could include this effect by testing the response of the system for both pipe position and string length.

Fig. 10.11. Marionette's motion for "Who Killed Cockrobin?" performed by actor 1.

The examples were limited to motions where the actor was told to stand in place during the performance. We could extend the range of feasible motions by adding more controllable strings and degrees of freedom. For example, a motor to control the string connecting the back would allow the marionette to bow. We could also add a pair of strings and motors to control the elbows independently or to move the entire marionette as a human operator would do for walking. In the construction of marionettes for human-operated performances extra strings are often added to enable a particular set of behaviors for that marionette's character.

We did not consider self collisions between the puppet and the strings or interaction with the environment. In the motions shown here we did not encounter situations where self collisions caused significant change of motion, but this issue is a serious concern in the design of performance marionettes with clothing that may catch on the strings. We kept the feet in contact with the floor to reduce the swing of the pelvis but did not explicitly consider contact with the environment in the control system. If the marionette had the additional degrees of freedom for such whole-body motions as walking, modeling of the interaction with the environment would be essential.

We explored two interfaces for driving the marionette: direct input of marker positions for realtime control and offline processing of human motion data. A third alternative would have been to capture a human-operated marionette performance to take advantage of the talent of a professional operator. The control scheme for this interface would presumably be significantly less complex because the motions would already be appropriate to the dynamics of the marionette. Such a system, however, could not easily be operated by an untrained user. In contrast, the control scheme described in this chapter enables a naive user to program a motorized marionette to create entertaining performances simply by performing the gestures in a motion capture system.

Part III

Conclusion

11

Conclusion and Future Work

11.1 Conclusion

This book presented new methods for dynamics computation and motion generation of human figures. These methods would serve as a computational foundation for practical applications of humanoids and CG characters, where human figures should move autonomously while interacting with human and/or the environment.

The dynamics computation methods can efficiently handle various phenomena observed in motions of human figures. The efficiency of basic dynamics and kinematics computation affects all phases of research and development activities on human figures, including motion analysis, dynamics simulation, motion generation, control, and mechanical design. The algorithms described in this book would therefore significantly accelerate various applications in both humanoid and CG animation.

The motion generation techniques focused on interactivity, which has seldom been discussed at motion or dynamics level. Interactive motion generation requires that the method accepts user inputs or online modification of the reference motion. Even if an intelligent behavior was generated through some interactions, the behavior is never embodied without a motion generator capable of making a physically consistent motion that fit the behavior. The methods presented in this book provide a framework for interactive and online motion generation considering physical consistency.

The two major contributions of this book are: (1) a computational foundation for realtime dynamics computation of structure-varying kinematic chains was established through the methods described in Chapters 2 to 6, and (2) a framework for interactive motion generator was developed in Chapters 7 to 10. The contributions of each chapter are summarized as follows:

(1) In Chapter 2, inverse and forward dynamics algorithms for general closed kinematic chains were presented. The methods are based on the efficient inverse dynamics computation for closed kinematic chains which projects the joint torques of the virtual open kinematic chain onto those of the actuated joints. The main contribution of this chapter is that it enabled the computation of the projection

matrix for general closed kinematic chains. It also automatically computes the degree of freedom and selects the generalized coordinates.

(2) Online and seamless dynamics computation of structure-varying kinematic chains was enabled by the link connectivity description that maintains the internal link connectivity data in accordance with the structural changes described in Chapter 3, in combination with the dynamics computation algorithms in Chapter 2. The description scheme adopts *pointers* to describe open kinematic chains and *virtual links* to describe closed ones. All link connections, even if the kinematic chain after the connection appears to be an open one, are handled by adding a virtual link, in order to avoid inversion of parent-child relationship.

(3) In Chapter 4, a parallel dynamics computation algorithm was presented. The algorithm achieves $O(\log N)$ asymptotic complexity with parallel computation on $O(N)$ processors for most practical kinematic chains including open and closed ones. The algorithm consists of two phases: the assembly phase to compute the temporary constraint forces and accelerations, and the disassembly phase to compute the final constraint forces and joint accelerations. Its major advantage is that the parallelism of the computation can be tuned for any kinematic chains by simply optimizing the order of processing the joints.

(4) In Chapter 5, a collision/contact model was presented. The model takes the advantage of rigid contact model to obtain stable results with large time steps. Unlike other rigid contact models, however, the method employs an iterative procedure to compute the contact forces and constraint conditions that satisfy the unilateral constraints to avoid time-consuming optimization computations.

(5) In Chapter 6, a soft collision/contact model was developed combining the efficient dynamics computation algorithm in Chapter 4 and implicit integration technique. An approximated implicit integration technique is applied to stabilize the simulation results with large spring and damper coefficients. The overall simulation system realizes nearly realtime dynamics simulation of human figure with collisions and contacts on an ordinary PC.

(6) In Chapter 7, an online motion generator called *dynamics filter* was proposed and implemented. Its basic function is to convert a physically inconsistent motion into consistent one. The physical consistency is ensured by searching for the optimal joint accelerations among the solutions of the equation of motion derived in Chapter 5. Since the dynamics filter does not require the whole reference motion in advance, we can interactively modify the reference motion during operation.

(7) In Chapter 8, a simple interface for generating whole-body motion from scratch called *pin-and-drag* interface was developed. The interface enables users to generate human-like motions without any reference motions by only constraining and dragging links. Although the method does not require pre-recorded motion data, it is also useful for online editing of existing motions. The enhanced inverse kinematics algorithm computes a motion that satisfies the constraints of the pin and dragged links as well as joint motion range and desired joint values at the same time. It yields reasonable results even if the human figure is in a singular

configuration or the constraints are inconsistent thanks to the singularity-robust inverse.

(8) In Chapter 9, the pin-and-drag interface was combined with the dynamics computation method in chapter 4 to generate physically consistent motions that satisfy the kinematic constraints input by the interface. The method makes use of the extended operational space inertia derived in the dynamics computation.

(9) In Chapter 10, the pin-and-drag interface was applied to controlling a motorized marionette using human motion capture data. The marker positions measured by an optical motion capture system were handled as multiple drag points. The original algorithm was extended by introducing two new classes of constraint to model the strings and gravity. We also dealt with the hardware-specific problems such as mapping the marker positions into the marionette's workspace and preventing swing.

Some of the algorithms described in this book are embedded in larger software packages. The forward dynamics algorithms in Chapters 2 and 4 are included in the simulation software package OpenHRP [20, 30, 66][1]. OpenHRP is developed for simulating motions of humanoid robots and has more than 100 users worldwide in the academics and industry. The enhanced inverse kinematics algorithm UTPoser, presented in Chapter 8, is used as the computational engine of Sega's award-winning animation software package Animanium^{TM} [39][2].

11.2 Future Work

The methods described in this book were mainly designed for whole-body motion of a single human figure composed of mechanical joints. We may extend these algorithms towards both larger and smaller scales: systems involving interactions between multiple human figures, and anatomical human models with bones, muscles, tendons, and organs.

The original motivation to develop the parallel dynamics computation scheme presented in Chapter 4 was to reduce the complexity to prevent the computational load from exploding when it is applied to large systems with hundreds of degrees of freedom. An example is a system with multiple human figures interacting with each others. It should be possible to apply the algorithms presented in this book to such systems without any problem. In addition, the effect of parallel processing would be even greater because they have a clear structural parallelism. For example, we can first handle each character or a group of characters by one process, and then use less number of processes for computing the interactions between them.

Another interesting extension is towards anatomical models of the human body. The first step would be to model the human body by bones, muscles, tendons, and organs based on anatomical data, but we may step further to the micro-scale structure of muscles. These models provide much better approximation of the human body.

[1] http://www.is.aist.go.jp/humanoid/openhrp/English/indexE.html

[2] http://www.animanium.com/

They can model extremely complex motions of real human joints [56], which are roughly approximated by spherical or rotational joints in common human figures. Each joint is also redundantly driven by a number of linear actuators rather than a single motor. Full dynamics computation of such models would open completely new fields in computer graphics, cognitive science, medicine and sport training, because we can directly compute the quantities sensed by the sensors on the real human body.

Usefulness of such models in medical field is rather obvious. For example, we can develop a rehabilitation program tailored to each person by building a model with the same disability. We may also be able to plan and optimize an orthopedic surgery by simulating the motions after several candidate strategies. In sport training, we can immediately point out the problems of a player by comparing his or her performance with an ideal one. These results might be of benefit to humanoid robotics field as well because we can study the control strategies of human motion and apply them to humanoid robots.

Anatomical human model is also essential for embodied cognitive science where the information from the real sensors plays a vital role. It is pointed out that the intelligence and the body of a creature are tightly connected with each other and cannot be separated [77]. In order to study human intelligence, therefore, we need a model with the same sensors and actuator as a real human. By studying the relationships between the sensor input and human mind, we might be able to quantitatively estimate the mental state of a human by only observing the motion because the sensor inputs can be simulated by the extended dynamics algorithms.

We can also find interesting applications in computer graphics. We would be able to render human figures with realistic muscle deformations using the force information computed by inverse dynamics computation. Some 3D CG software packages [14] do provide a similar function but they only use kinematic information. The forward dynamics computation would yield more realistic motions, although controlling such models will be extremely difficult.

We have just stepped the first step towards the modeling and dynamics computation of anatomical human figures. We build a musculo-skeletal model with 155 DOF driven by more than 500 muscles, tendons, and ligaments. Fig. 11.1 visualizes the results of inverse dynamics computation of a kicking motion [68]. The color of each muscle changes from light color (yellow for readers with color information) to dark color (red) as the force required to perform the motion increases. We used the string constraints added in Chapter 10 for the tendons and ligaments and applied the inverse kinematics algorithm to computing the joint angles from motion capture data. The joint forces computed by the inverse dynamics are distributed among the muscles by a numerical optimization techniques because the joints are redundantly actuated and the muscles can only pull the bones.

Currently anatomical human models are usually studied from the medical or biomechanical point of view, thus using biological information to compute the muscle forces. We believe that the approach from the robotics field provides useful and efficient tools for analyzing and simulating motions of such models.

Another interesting extension is the closer connection with motion capture data. Although the examples in Chapters 7 and 8 partially use motion capture data, the

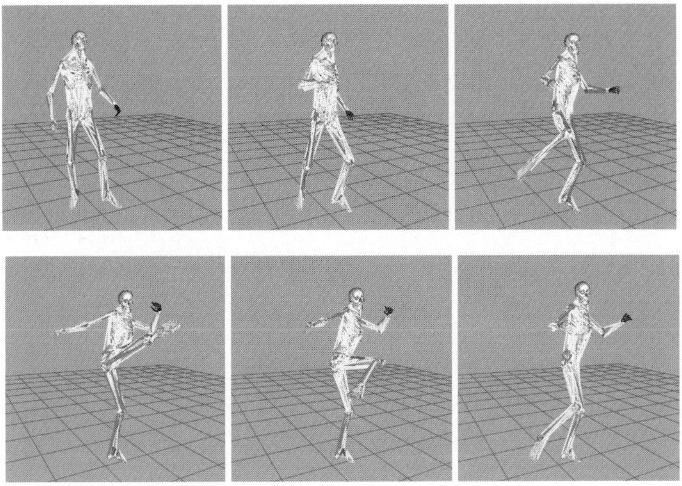

Fig. 11.1. Visualization of the muscle forces for a kick motion.

algorithms are basically model-based approaches. A recent trend in computer animation is the data-based synthesis from motion capture data because it is relatively easier to obtain human-like motions. One of the popular approaches is to first extract the possible transitions among postures from a long motion clip, and then search for a sequence of transitions that best fits the higher-level commands from the user [4, 46, 44]. However, heavily relying on the existence of motion capture data may fail if, for example, the user required a completely different motion. Pure model-based methods, on the other hand, usually suffer from difficulty of interactive control as illustrated in Chapters 7 and 9. Therefore, a clever combination of data-based and model-based approaches seems to be the right solution.

References

1. F. M. Amirouche, *Computational Methods in Multibody Dynamics*. Upper Saddle River, NJ: Prentice Hall, 1992.
2. E. Anderson, Z. Bai, C. Bischof, S. Blackford, J. Demmel, J. Dongarra, J. Du Croz, A. Greenbaum, S. Hammarling, A. McKenney, and D. Sorensen, *LAPACK Users' Guide*, 3rd ed. Philadelphia, PA: Society for Industrial and Applied Mathematics, 1999.
3. K. Anderson and S. Duan, "Highly Parallelizable Low-Order Dynamics Simulation Algorithm for Multi-Rigid-Body Systems," *AIAA Journal on Guidance, Control and Dynamics*, vol. 23, no. 2, pp. 355–364, 2000.
4. O. Arikan and D. A. Forsyth, "Synthesizing Constrained Motions from Examples," *ACM Transactions on Graphics*, vol. 21, no. 3, pp. 483–490, 2002.
5. N. Badler, K. Manoochehri, and D. Baraff, "Multi-dimensional Input Techniques and Articulated Figure Positioning by Multiple Constraints," in *Proceedings of the 1986 Workshop on Interactive 3D Graphics*, 1986, pp. 151–169.
6. N. Badler, C. Phillips, and B. Webber, *Simulating Humans*. New York, NY: Oxford University Press, 1993.
7. D. Bae and E. Haug, "A Recursive Formulation for Constrained Mechanical System Dynamics: PartI. Open Loop Systems," *Mechanics of Structures and Machines*, vol. 15, no. 3, pp. 359–382, 1987.
8. ——, "A Recursive Formulation for Constrained Mechanical System Dynamics: PartII. Closed Loop Systems," *Mechanics of Structures and Machines*, vol. 15, no. 4, pp. 481–506, 1987-88.
9. D. Baraff, "Coping with Friction for Non-penetrating Rigid Body Simulation," in *Proceedings of SIGGRAPH '91*, 1991, pp. 31–40.
10. ——, "Fast Contact Force Computation for Nonpenetrating Rigid Bodies," in *Proceedings of SIGGRAPH '94*, 1994, pp. 23–34.
11. D. Baraff and A. Witkin, "Large Steps in Cloth Simulation," in *Proceedings of SIGGRAPH '98*, 1998, pp. 43–54.
12. D. Baraff, "Linear-Time Dynamics Using Lagrange Multipliers," in *Proceedings of SIGGRAPH '96*, 1996, pp. 137–146.
13. D. Brogan, R. Metoyer, and J. Hodgins, "Dynamically Simulated Characters in Virtual Environments," *IEEE Computer Graphics and Applications*, vol. 18, no. 5, pp. 58–69, 1998.
14. CGCharacter, http://www.cgcharacter.com/.

15. K. Chang and O. Khatib, "Efficient Algorithm for Extended Operational Space Inertia Matrix," in *Proceedings of IEEE International Conference on Robotics and Automation*, 1999, pp. 350–355.
16. ——, "Operational Space Dynamics: Efficient Algorithms for Modeling and Control of Branching Mechanisms," in *Proceedings of IEEE International Conference on Robotics and Automation*, 2000, pp. 850–856.
17. K. Choi and H. Ko, "Online Motion Retargetting," *The Journal of Visualization and Computer Animation*, vol. 11, pp. 223–235, 2000.
18. J. Craig, *Introduction to Robotics: Mechanics and Control.* Reading, MA: Addison-Wesley, 1986.
19. A. DasGupta and Y. Nakamura, "Making Feasible Walking Motion of Humanoid Robots from Human Motion Captured Data," in *Proceedings of International Conference on Robotics and Automation*, 1999, pp. 1044–1049.
20. F. Kanehiro et. al, "Virtual humanoid robot platform to develop controllers of real humanoid robots without porting," in *Proceedings of IEEE/RSJ International Conference on Intelligent Robots and Systems*, 2001, pp. 1093–1099.
21. A. Faloutsos, M. van de Panne, and D. Terzopoulos, "Composable Controllers for Physics-Based Character Animation," in *Proceedings of SIGGRAPH 2001*, 2001, pp. 251–260.
22. R. Featherstone, *Robot Dynamics Algorithm.* Boston, MA: Kluwer Academic Publishers, 1987.
23. ——, "A Divide-and-Conquer Articulated-Body Algorithm for Parallel $O(\log(n))$ Calculation of Rigid-Body Dynamics. Part1: Basic Algorithm," *International Journal of Robotics Research*, vol. 18, no. 9, pp. 867–875, 1999.
24. ——, "A Divide-and-Conquer Articulated-Body Algorithm for Parallel $O(\log(n))$ Calculation of Rigid-Body Dynamics. Part2: Trees, Loops, and Accuracy," *International Journal of Robotics Research*, vol. 18, no. 9, pp. 876–892, 1999.
25. R. Featherstone and A. Fijany, "A Technique for Analyzing Constrained Rigid-Body Systems, and its Application to the Constraint Force Algorithm," *IEEE Transactions on Robotics and Automation*, vol. 15, no. 6, pp. 1140–1144, 1999.
26. A. Fijany, I. Sharf, and G. D'Eleuterio, "Parallel $O(\log N)$ Algorithms for Computation of Manipulator Forward Dynamics," *IEEE Transactions on Robotics and Automation*, vol. 11, no. 3, pp. 389–400, 1995.
27. Y. Fujimoto, S. Obata, and A. Kawamura, "Robust Biped Walking with Active Interaction Control between Foot and Ground," in *Proceedings of International Conference on Robotics and Automation*, 1998, pp. 2030–2035.
28. M. Gleicher, "Retargetting Motion to New Characters," in *Proceedings of SIGGRAPH '98*, 1998, pp. 33–42.
29. S. Gottschalk, M. Lin, and D. Manocha, "OBB-Tree: A Hierarchical Structure for Rapid Interference Detection," in *Proceedings of SIGGRAPH '96*, 1996, pp. 171–180.
30. H. Hirukawa et. al, "Open architecture humanoid robot platform," in *Proceedings of International Symposium on Robotics Research*, 2001.
31. E. Haug, *Computer Aided Kinematics and Dynamics of Mechanical Systems.* Needham Heights, MA: Allyn and Bacon Series in Engineering, 1989.
32. J. Hodgins, W. Wooten, D. Brogan, and J. O'Brien, "Animating Human Athletics," in *Proceedings of ACM SIGGRAPH '95*, 1995, pp. 71–78.
33. G. Hoffmann, "Teach-In of a Robot by Showing the Motion," in *IEEE International Conference on Image Processing*, 1996, pp. 529–532.
34. M. Hollars, D. Rosenthal, and M. Sherman, "SD-Fast," Symbolic Dynamics, Inc., 1991.

35. "The Honda Humanoid Robot ASIMO," http://world.honda.com/ASIMO/.
36. J. Hu, J. Pratt, and G. Pratt, "Adaptive Dynamic Control of a Biped Walking Robot with Radial Basis Function Neural Networks," in *Proceedings of International Conference on Robotics and Automation*, 1998, pp. 400–405.
37. Q. Huang, K. Kaneko, K. Yokoi, S. Kajita, and T. Kotoku, "Balance Control of a Biped Robot Combining Off-line Pattern with Real-time Modification," in *Proceedings of International Conference on Robotics and Automation*, 2000, pp. 3346–3352.
38. K. Hunt and F. Crossley, "Coefficient of Restitution Interpreted as Damping in Vibroimpact," *ASME Journal of Applied Mechanics*, pp. 440–445, 1988.
39. H. Imagawa, "The Animator-Oriented Motion Generator Animanium Based on a Humanoid Robots Control Algorithm," in *Sketches and Applications, SIGGRAPH 2001*, 2001.
40. S. Kajita and K. Tani, "Experimental Study of Biped Dynamic Walking in the Linear Inverted Pendulum Mode," in *Proceedings of the IEEE International Conference on Robotics and Automation*, 1995, pp. 2885–2891.
41. Y. Kim, M. Otaduy, M. Lin, and D. Manocha, "Fast Penetration Depth Computation for Physically-based Animation," in *ACM Symposium on Computer Animation*, 2002, pp. 23–31.
42. J. Kleinfinger and W. Khalil, "Dynamic Modeling of Closed-Loop Robots," in *Proceedings of the 16th International Symposium on Industrial Robots*, 1986, pp. 401–412.
43. H. Ko and N. Badler, "Animating Human Locomotion with Inverse Dynamics," *IEEE Transactions on Computer Graphics*, vol. 16, no. 2, pp. 50–59, 1996.
44. L. Kovar, M. Gleicher, and F. Pighin, "Motion graphs," in *Proceedings of SIGGRAPH 2002*, 2002, pp. 473–482.
45. Y. Kuroki, T. Ishida, J. Yamaguchi, M. Fujita, and T. Doi1, "A Small Biped Entertainment Robot," in *Proceedings of Humanoids 2001*, 2001.
46. J. Lee, J. Chai, P. S. A. Reitsma, J. K. Hodgins, and N. S. Pollard, "Interactive Control of Avatars Animated With Human Motion Data," *ACM Transactions on Graphics*, vol. 21, no. 3, pp. 491–500, 2002.
47. J. Lee and S. Shin, "A Hierarchical Approach to Interactive Motion Editing for Human-like Figures," in *Proceedings of SIGGRAPH '99*, 1999, pp. 39–48.
48. K. Lilly and D. Orin, "Efficient O(n) Recursive Computation of the Operational Space Inertia Matrix," *IEEE Transactions on Systems, Man, and Cybernetics*, vol. 23, no. 5, pp. 1384–1391, 1993.
49. L. Ljung, *System Identification – Theory for the User*. Upper Saddle River, NJ: Prentice Hall, 1987.
50. P. Lötstedt, "Numerical Simulation of Time-Dependent Contact Friction Problems in Rigid Body Mechanics," *SIAM Journal of Scientific Statistical Computing*, vol. 5, no. 2, pp. 370–393, 1984.
51. C. Lubich, U. Nowak, U. Poehle, and C. Engstler, *MEXX – Numerical Software for the Integration of Constrained Mechanical Multibody Systems*. ftp://na.uni-tuebingen.de/pub/engstler/mexx/SC-92-12.ps.gz, 1992.
52. J. Luh, M. Walker, and R. Paul, "On-line Computational Scheme for Mechanical Manipulators," *ASME Journal on Dynamic Systems, Measurment and Control*, vol. 104, pp. 69–76, 1980.
53. ——, "Resolved Acceleration Control of Mechanical Manipulators," *IEEE Transactions on Automatic Control*, vol. 25, no. 3, pp. 468–474, 1980.
54. A. Maciejewski, "Dealing with the Ill-conditioned Equations of Motion for Articulated Figures," *IEEE Computer Graphics and Applications*, vol. 10, no. 3, pp. 63–71, 1990.

55. D. Marhefka and D. Orin, "Simulation of Contact Using a Nonlinear Damping Model," in *Proceedings of the 1996 IEEE International Conference on Robotics and Automation*, 1996, pp. 1662–1668.
56. W. Maurel and D. Thalmann, "Human Shoulder Modeling Including Scapulo-Thoracic Constraint and Joint Sinus Cones," *Computers and Graphics*, vol. 24, pp. 203–218, 2000.
57. S. McMillan and D. Orin, "Forward Dynamics of Multilegged Vehicles Using the Composite Rigid Body Method," in *Proceedings of IEEE International Conference on Robotics and Automation*, 1998, pp. 464–470.
58. J. McPhee, "Automatic Generation of Motion Equations for Planar Mechanical Systems Using the New Set of "Branch Coordinates"," *Mechanism and Machine Theory*, vol. 33, no. 6, pp. 805–823, 1998.
59. B. V. Mirtich, "Impulse-Based Dynamic Simulation of Rigid Body Systems," Ph.D. dissertation, University of California, Berkeley, 1996.
60. N. M'sirdi, N. Manamani, and N. Nadjar-Gauthier, "Methodology based on CLC for Control of Fast Legged Robots," in *Proceedings of the IEEE/RSJ International Conference on Intelligent Robotics and Systems*, 1998, pp. 71–76.
61. K. Nagasaka, I. Masayuki, and H. Inoue, "Walking Pattern Generation for a Humanoid Robot Based on Optimal Gradient Method," in *Proceedings of IEEE International Conference on Systems, Man, and Cybernetics*, 1999, pp. 908–913.
62. Y. Nakamura, *Advanced Robotics—Redundancy and Optimization*. Reading, MA: Addison–Wesley , 1991.
63. Y. Nakamura and M. Ghodoussi, "Dynamics Computation of Closed-Link Robot Mechanisms with Nonredundant and Redundant Actuators," *IEEE Transactions on Robotics and Automation*, vol. 5, no. 3, pp. 294–302, 1989.
64. Y. Nakamura and H. Hanafusa, "Inverse Kinematics Solutions with Singularity Robustness for Robot Manipulator Control," *Journal of Dynamic Systems, Measurement, and Control*, vol. 108, pp. 163–171, 1986.
65. ——, "Task Priority Based Redundancy Control of Robot Manipulators," *International Journal of Robotics Research*, vol. 6, no. 2, pp. 3–15, 1987.
66. Y. Nakamura, H. Hirukawa, and K. Yamane et al., "Humanoid Robot Simulator for the METI HRP Project," *Robotics and Autonomous Systems*, vol. 37, pp. 101–114, 2001.
67. Y. Nakamura and T. Ropponen, "Actuation Redundancy of a Closed Link Manipulator," in *Proceedings of 1990 American Control Conference*, 1990, pp. 2294–2299.
68. Y. Nakamura, K. Yamane, I. Suzuki, and Y. Fujita, "Dynamic Computation of Musculo-Skeletal Human Model Based on Efficient Algorithm for Closed Kinematic Chains," in *Proceedings of International Symposium on Adaptive Motion of Animals and Machines*, 2003, pp. SaP–I–2.
69. Y. Nakamura, Y. Yokokohji, H. Hanafusa, and T. Yoshikawa, "Unified Recursive Formulation of Kinematics and Dynamics for Robot Manipulators," in *Proceedings of Japan-USA Symposium on Flexible Automation*, 1986, pp. 53–60.
70. K. Nishiwaki, T. Sugihara, S. Kagami, M. Inaba, and H. Inoue, "Online Mixture and Connection of Basic Motions for Humanoid Walking Control by Footprint Specification," in *Proceedings of the IEEE International Conference on Robotics and Automation*, 2001, pp. 4110–4115.
71. D. Orin, R. McGhee, M. Vukobratovic, and G. Hartoch, "Kinematic and Kinetic Analysis of Open-chain Linkages Utilizing Newton-Euler Methods," *Mathematical Biosciences*, vol. 43, pp. 107–130, 1979.
72. D. Orin and W. Schrader, "Efficient Computation of the Jacobian for Robot Manipulators," *International Journal of Robotics Research*, vol. 3, no. 4, pp. 66–75, 1984.

73. Parallel and Distributed System Software Laboratory, *SCore*. http://pdswww.rwcp.or.jp/home.html, 2001.
74. J. Park and K. Kim, "Biped Robot Walking Using Gravity-Compensated Inverted Pendulum Mode and Computed Torque Control," in *Proceedings of the IEEE International Conference on Robotics and Automation*, 1998, pp. 3528–3533.
75. J. Park and Y. Rhee, "ZMP Trajectory Generation for Reduced Trunk Motions of Biped Robots," in *Proceedings of the 1998 IEEE/RSJ International Conference on Intelligent Robots and Systems*, 1998, pp. 90–95.
76. B. Perrin, C. Chevallereau, and C. Verdier, "Calculation of the Direct Dynamics Model of Walking Robots: Comparison Between Two Methods," in *Proceedings of IEEE International Conference on Robotics and Automation*, 1997, pp. 1088–1093.
77. R. Pfeifer and C. Scheier, *Understanding Intelligence*. Cambridge, MA: MIT Press, 2001.
78. F. Pfeiffer and C. Glocker, *Multibody Dynamics with Unilateral Contacts*. Wiley Series in Nonlinear Science, 1996.
79. C. Phillips and N. Badler, "Jack: A Toolkit for Manipulating Articulated Figures," in *Proceedings of ACM/SIGGRAPH Symposium on User Interface Software*, 1988, pp.221–229.
80. C. Phillips, J. Zhao, and N. Badler, "Interactive Real-time Articulated Figure Manipulation Using Multiple Kinematic Constraints," in *Proceedings of the 1990 Workshop on Interactive 3D Graphics*, 1990, pp. 245–250.
81. N. Pollard and F. Behmaram-Mosavat, "Force-Based Motion Editing for Locomotion Tasks," in *Proceedings of IEEE International Conference on Robotics and Automation*, 2000, pp. 663–669.
82. N. Pollard, J. Hodgins, M. Riley, and C. Atkeson, "Adapting Human Motion for the Control of a Humanoid Robot," in *Proceedings of the IEEE International Conference on Robotics and Automation*, 2002, pp. 1390–1397.
83. Z. Popović, "Editing Dynamic Properties of Captured Human Motion," in *Proceedings of IEEE International Conference on Robotics and Automation*, 2000, pp. 670–675.
84. W. Press, S. Teukolsky, W. Vetterling, and B. Flannery, *Numerical Recipes in C Second Edition*. Cambridge, UK: Cambridge University Press, 1999.
85. C. Rahn, F. Zhang, S. Joshi, and D. Dawson, "Asymptotically stabilizing angle feedback for a flexible cable gantry crane," *ASME Journal of Dynamic Systems, Measurement, and Control*, vol. 121, pp. 563–566, 1999.
86. C. Rose, M. Cohen, and B. Bodenheimer, "Verbs and Adverbs: Multidimentional Motion Interpolation," *IEEE Computer Graphics and Applications*, vol. 18, no. 5, pp. 32–40, 1998.
87. C. Rose, B. Guenter, B. Bodenheimer, and M. Cohen, "Efficient Generation of Motion Transitions using Spacetime Constraints," in *In Proceedings of SIGGRAPH '96*, 1996, pp. 147–154.
88. C. Rose, P.-P. Sloan, and M. Choen, "Artist-Directed Inverse-Kinematics Using Radial Basis Function Interpolation," *Eurographics*, vol. 20, no. 3, 2001.
89. D. Rosenthal, "An Order n Formulation for Robotic Systems," *The Journal of the Astronautical Sciences*, vol. 38, no. 4, pp. 511–529, 1990.
90. "Sarcos High Performance Robots," http://www.sarcos.com/entspec_highperfrobot.html.
91. D. Stewart and J. Trinkle, "An Implicit Time-Stepping Scheme for Rigid Body Dynamics with Coulomb Friction," in *Proceedings of IEEE International Conference on Robotics and Automation*, 2000, pp. 162–169.

92. T. Sugihara, Y. Nakamura, and H. Inoue, "Realtime Humanoid Motion Generation through ZMP Manipulation based on Inverted Pendulum Control," in *Proceedings of the IEEE International Conference on Robotics and Automation*, 2002, pp. 1404–1409.
93. D. Surla and M. Rackovic, "Closed-Form Mathematical Model of the Dynamics of Anthropomorphic Locomotion Robotic Mechanism," in *2nd ECPD International Conference on Advanced Robotics, Intelligent Automation and Active Systems*, 1996, pp. 327–331.
94. S. Tak, O. Song, and H. Ko, "Motion balance filtering," *Eurographics 2000, Computer Graphics Forum*, vol. 19, no. 3, pp. 437–446, 2000.
95. M. van de Panne, "Towards Agile Animated Characters," in *Proceedings of IEEE International Conference on Robotics and Automation*, 2000, pp. 682–687.
96. M. Walker and D. Orin, "Efficient Dynamic Computer Simulation of Robot Manipulators," *ASME Journal on Dynamic Systems, Measurement and Control*, vol. 104, pp. 205–211, 1982.
97. W. Wooten and J. Hodgins, "Animation of Human Diving," *Computer Graphics Forum*, vol. 15, no. 1, pp. 3–13, 1996.
98. J. Yamaguchi, A. Takanishi, and I. Kato, "Development of a Biped Walking Robot Compensating for Three-axis Moment by Trunk Motion," in *Proceedings of the IEEE/RSJ International Conference on Intelligent Robotics and Systems*, 1993, pp. 561–566.

Index

Springer Tracts in Advanced Robotics

Edited by B. Siciliano, O. Khatib, and F. Groen
Published Titles:

Vol. 1: Caccavale, F.; Villani, L. (Eds.)
Fault Diagnosis and Fault Tolerance for Mechatronic Systems: Recent Advances
191 pp. 2002 [3-540-44159-X]

Vol. 2: Antonelli, G.
Underwater Robots: Motion and Force Control of Vehicle-Manipulator Systems
209 pp. 2003 [3-540-00054-2]

Vol. 3: Natale, C.
Interaction Control of Robot Manipulators: Six-degrees-of-freedom Tasks
120 pp. 2003 [3-540-00159-X]

Vol. 4: Bicchi, A.; Christensen, H.I.; Prattichizzo, D. (Eds.)
Control Problems in Robotics
296 pp. 2003 [3-540-00251-0]

Vol. 5: Siciliano, B.; Dario, P. (Eds.)
Experimental Robotics VIII
685 pp. 2003 [3-540-00305-3]

Vol. 6: Jarvis, R.A.; Zelinsky, A. (Eds.)
Robotics Research – The Tenth International Symposium
580 pp. 2003 [3-540-00550-1]

Vol. 7: Boissonnat, J.-D.; Burdick, J.; Goldberg, K.; Hutchinson, S. (Eds.)
Algorithmic Foundations of Robotics V
577 pp. 2004 [3-540-40476-7]

Vol. 8: Baeten, J.; De Schutter, J.
Integrated Visual Servoing and Force Control
198 pp. 2004 [3-540-40475-9]

Printing: Mercedes-Druck, Berlin
Binding: Stein + Lehmann, Berlin